Praise for *The Dressmaker's Mirror: Sudden Death, Genetics, and a Jewish Family's Secret*

"*The Dressmaker's Mirror* is a remarkable book. Susan Weiss Liebman's many life roles coalesce in this humane defense of the importance of knowledge when confronting life events that we cannot hope to understand or fully control. This book will be of immediate interest to young women with special talents, members of families affected by genetic disease, Americans appalled by resurgent antisemitism, and other definable groups. More expansively, it will appeal to anyone looking for a captivating story about life lived large." —**Mayard Olson**, PhD, one of the founders of the Human Genome Project; Gruber Prize in Genetics

"Dr. Susan Weiss Liebman, an accomplished geneticist herself, gives a non-technical introduction to the current state of this field, embedded in her own family story contending with one such condition. The special social/emotional implications of genetic diseases are vividly displayed in these personal recollections, and the current and emerging approaches to diagnosis and therapy ably described in an accessible way." —**Reed Wickner**, MD, PhD, NIH Distinguished investigator; Elected to US National Academy of Sciences

"I love *The Dressmaker's Mirror*. Dr. Liebman has worked tirelessly to help shed light on this devastating disease and we applaud her for her efforts." —**Greg Ruf**, Founder and Executive Director of the Dilated Cardiomyopathy Foundation

"Dr. Susan Weiss Liebman has written a fascinating discovery of an unknown mutation affecting the heart that is unique to the Ashkenazi Jewish population." —**James Haber**, PhD, elected to US National Academy of Sciences and Director; Rosenstiel Basic Medical Sciences Research Center, Brandeis University

"This book is a beautiful example of how an understanding of human genetics and human genomics informs our understanding of the past, the present, and our likely future." —**Thomas Petes**, PhD, Minnie Geller Distinguished Professor of Research in Genetics, Duke School of Medicine; elected to US National Academy of Sciences

The Dressmaker's Mirror

Sudden Death, Genetics, and a Jewish Family's Secret

SUSAN WEISS LIEBMAN, PHD

ROWMAN & LITTLEFIELD
Lanham • Boulder • New York • London

Published by Rowman & Littlefield
An imprint of The Rowman & Littlefield Publishing Group, Inc.
4501 Forbes Boulevard, Suite 200, Lanham, Maryland 20706
www.rowman.com

86-90 Paul Street, London EC2A 4NE

Distributed by NATIONAL BOOK NETWORK

Copyright © 2024 by Susan Weiss Liebman

All rights reserved. No part of this book may be reproduced in any form or by any electronic or mechanical means, including information storage and retrieval systems, without written permission from the publisher, except by a reviewer who may quote passages in a review.

British Library Cataloguing in Publication Information Available

Library of Congress Cataloging-in-Publication Data

Names: Liebman, Susan Weiss, 1947- author.
Title: The dressmaker's mirror : sudden death, genetics, and a Jewish family's secret / Susan Weiss Liebman, PhD.
Other titles: Sudden death, genetics, and a Jewish family's secret
Description: Lanham : Rowman & Littlefield Publishers, [2024]. | Includes bibliographical references.
Identifiers: LCCN 2024030990 (print) | LCCN 2024030991 (ebook) | ISBN 9781538196809 (cloth) | ISBN 9781538196816 (epub)
Subjects: LCSH: Ashkenazim—United States—Diseases—Genetic aspects. | Heart—Diseases—Genetic aspects. | Myocardium—Diseases. | Liebman, Susan Weiss, 1947- —Family. | Weiss family. | Medical genetics. | Jews—United States—Diseases—Genetic aspects.
Classification: LCC RB155.5 .L538 2024 (print) | LCC RB155.5 (ebook) | DDC 616.042—dc23/eng/20240708
LC record available at https://lccn.loc.gov/2024030990
LC ebook record available at https://lccn.loc.gov/2024030991

∞™ The paper used in this publication meets the minimum requirements of American National Standard for Information Sciences—Permanence of Paper for Printed Library Materials, ANSI/NISO Z39.48-1992

This book is in loving memory of my niece Karen Stephanie Rothman Fried, her unborn son James Alex, my sister Diane Weiss Rothman, and my parents Norman and Cyrilla Weiss. It is dedicated to the future, in the capable hands of my four grandchildren, Aaron, Amy, Adam, and Mark.

Contents

Preface		ix
1	The Dressmaker's Mirror	1
2	Catastrophe	13
3	Interlude: Genetic Testing of Infants	21
4	Family Life	27
5	Self-Discovery	45
6	The Packed Bags	57
7	Coming of Age	67
8	The Dormitory	81
9	The Rabbit	89
10	Her Memory Is a Blessing	101
11	Becoming a Geneticist	109
12	Professor Sue	123

CONTENTS

13	Interlude: Pros and Cons of Genetic Testing and Screening	131
14	Loss of Innocence	139
15	The Aftermath	149
16	The Mitzvah	159
17	Interlude: The Central Dogma of Molecular Biology and Me	165
18	Empty Nests and Full Hearts	171
19	Facing Fate	185
20	Hunting the Killer	195
21	The Dressmaker's Secret	209

Advocacy for Genetic Testing	221
Acknowledgments	227
Appendix 1: "Dear Family" Letter	229
Appendix 2: Genetic Testing Sources	233
Notes	237
Glossaries	247
Bibliography	255
About the Author	263

Preface

This book explains how to find out if you or a loved one have a mutation likely to cause treatable heart, cancer, or other disease *before* it shatters your family. I believe I became a geneticist, at a time when few women pursued this path, because I was destined to help understand the family illness and advocate for genetic screening. My niece's heart stopped beating one day when she was thirty-six and in her prime. Her autopsy report brought other family illnesses and early deaths into focus. Taken together, this caused the family to panic; there seemed to be a genetic problem that could trigger sudden death at any time. As a mother, I prayed for the curse to spare my children. As a geneticist, I set out to find the killer. Along the way, I became an expert in the burgeoning field of genetic testing.

In this book, I tell of the grief my family faced for generations because of this affliction, how we found the mutation behind it, and how finding it transformed our lives. The stories are true, although memory is imperfect, and I have changed a few names and details to protect people's privacy. The dialogue conveys the sense of the conversations, although I don't know the exact words that people used.

I pepper these personal stories with a few general discussions of genetic testing, screening, and gene therapy. I use the word "mutation," because it is easier for a general audience to understand than the term "variant" (pathogenic or benign) that has come to replace it.

Both the American Cancer Society (ACS) and the American Heart Association (AHA) recommend genetic testing for those with a family history of disease.[1] The AHA advises testing even without family history if a patient has a condition commonly caused by a genetic mutation. Knowing the disease mutation lets other family members check if they have it and get preventive care. Unfortunately, despite widespread agreement among experts on the value of such testing, the clinic often fails to implement it.[2]

My family story has driven me to promote genetic screening in addition to genetic testing. I posit that proactive screens for mutations that cause treatable medical conditions regardless of family history would save lives and improve health outcomes. Such tests are now quite cost-effective. However, ClinGen is concerned that we neither have enough physicians who are sufficiently trained nor the follow-up clinical resources required.[3] Also, the American College of Medical Genetics and Genomics asserts more research is needed to understand how the risks of certain disease-causing mutations vary in different populations.[4] Another concern is that knowing you have a higher risk of illness may cause emotional distress even if treatment is available. Also, if revealed, genetic results could lead to social and employment discrimination and the denial of life insurance.

I hope my family's story will inspire more action toward tackling these serious issues and will raise awareness of the value of genetic testing and screening.

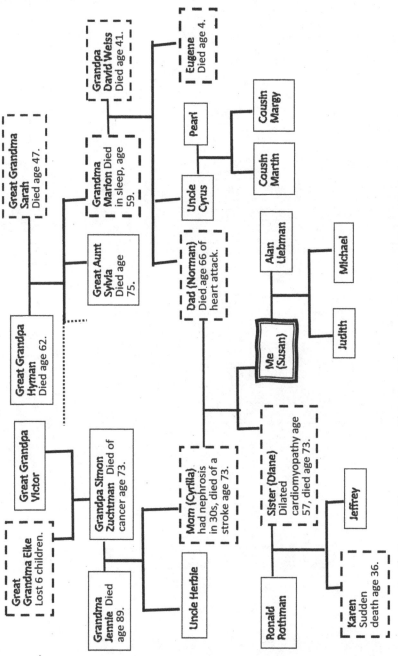

Family Tree. Family members with unusual illness or early death are boxed in dashes. AUTHOR CREATED

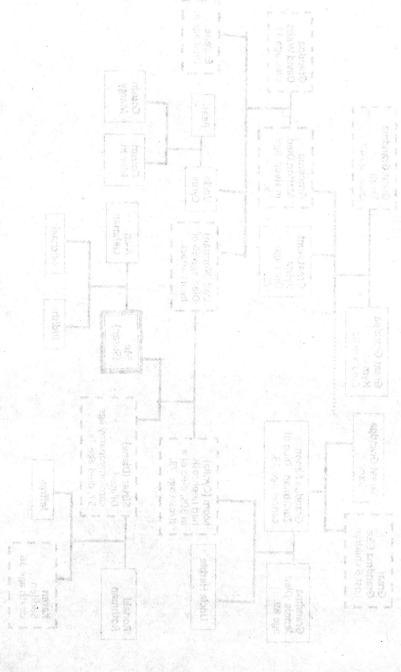

ONE

The Dressmaker's Mirror

> "On Rosh Hashanah it is inscribed, and on Yom Kippur it is sealed ... who shall live and who shall die, who in good time, and who by an untimely death ..." —Traditional Hebrew Prayer

Everything was different now. I just didn't know it yet. *Ring, ring, ring!* I stopped setting the dinner table and picked up the telephone receiver. It was my sister's son, Jeffrey. His voice sounded weird—too deep, too matter of fact, and the traditional "Hi" was missing. "Susan, it's Jeffrey. Karen died."

Just a few hours ago my older sister Diane had gushed with me on the phone about remodeling her daughter Karen's Brooklyn high-rise co-op to include a nursery. Karen was only thirty-six, pregnant with her first child. She couldn't have died. I was angry. Unable to take Jeffrey's news in, I responded with my first thought. "Is this some sort of joke?"

My question didn't insult or surprise him. He focused on getting me to accept his news. "You know I wouldn't joke about this."

His words made no sense. My knees collapsed and I landed on a chair, still clutching the phone. "You mean her baby died."

CHAPTER ONE

"Karen died and the baby died with her."

My face contorted. Words stuck in my throat. "What happened?"

"She was at a restaurant with Andrew when she just collapsed." Karen and her husband, Andrew, lived in Brooklyn, while my nephew Jeffrey lived near his parents' winter home in Florida. "Bystanders used Andrew's cellphone to reach his sister. She got to the hospital in time to support Andrew when doctors pronounced Karen dead. That's when she called me."

How could this have happened? Young healthy people don't just die. Was it poison? Murder? Drugs? My body reacted with cramps, its usual response to panic. Gasping, I struggled to spit out each word. "Just died for no reason? How is that possible?" Did something go wrong with the pregnancy? I couldn't help but think about my own pregnant daughter—was she in danger?

"We're going to the airport now. Mom wants to see Karen tonight. She says for you to come to her Brooklyn co-op right away. She wants to be with you."

"Of course." How will Diane survive this? How can I comfort her?

I heard my nephew's bass voice, "I have to go," along with wailing in the background. They were rushing to catch their flight. Then came the dial tone.

I looked at the phone in disbelief. The receiver fell from my hand dangling on the cord as I ran yelling for my husband.

He raced down the stairs. "What's wrong?"

"Karen and her baby died. I can't believe it. She just collapsed and died."

I rocked back and forth on the floor.

Alan took me in his arms and held me. "I'm so sorry." He booked me on an early morning flight out of Chicago. He would follow the next day as the funeral would be a couple of days later.

Between bouts of crippling cramps and diarrhea, I contacted a colleague at the university to teach my genetics course and threw black clothes in a suitcase. My flight was on time. Alan must have told them it was a bereavement trip because they seated me in first class.

The plane shuddered as it climbed. I stared out my window as white mist enveloped the plane. It was November 16, 2008, the twenty-eighth anniversary of my father's deadly heart attack. The mist helped me imagine my father shaking his head as Karen joined him in heaven, saying, "You weren't supposed to come yet."

By the time of Karen's death, I had been a genetics professor and researcher for over thirty years. I had always considered genetics in terms of my career. I studied it as an undergraduate at the Massachusetts Institute of Technology (MIT) and a graduate student and postdoctoral fellow at Harvard and the University of Rochester. I understood that genetics determined my eye and hair color, my handedness, and other traits, but I never thought that it would have a significant impact on my actual life. Now that all changed. I was terrified that history would repeat itself with the death of my pregnant daughter.

This wasn't the first time someone in my family died suddenly, but I was determined to make it the last.

I learned about another sudden death the year Watson and Crick first described DNA as a tiny, twisted Jacob's ladder descending from heaven. One evening, when I was six, my father read to me as we lounged on his bed leaning back on a curved wooden headboard. This time was precious, because his two full-time jobs—as a high school teacher and a print shop owner—left little time for me. I felt secure and loved with his arm around my shoulder. The family often gathered here in my parents' bedroom facing Ocean Avenue, a bustling and noisy thoroughfare in the Flatbush neighborhood of Brooklyn.

CHAPTER ONE

As Dad read, his voice cracked and his hazel gray eyes teared. This was his normal response when reading sentimental children's books.

My mother also reclined on her twin bed that night, leaning on the mahogany headboard that her decorator cousin picked out for us. Mom listened to the radio softly recount the news. Dad affectionately called her his newshound.

On top of my parents' double dresser sat a framed sepia proof of my mother's bridal portrait, and an oval mirror stand displaying fancy perfume bottles. Mom let my big sister Diane and me rummage through the dresser, whose upper drawers yielded scissors, paper clips, and other useful items. She often warned us to be careful around her perfume bottles, as they were her pride and joy. A thick, clear glass slab protected the top of the dresser. I loved to study the black-and-white family photographs that lay scattered beneath the glass.

After reading the book, I wandered across the flat green carpet to look at the photos. I pointed to a small snapshot of a child I didn't recognize. "Who's that little boy?"

Dad pursed his lips and paused before he answered. "That's my brother."

This confused me because my father had only one brother, Cyrus, and he didn't look like the child in the photo. To be sure, I tilted my head and asked, "You mean Uncle Cy?"

Dad opened his arms with his palms up. "No, another brother."

This was big. I had another uncle. But why didn't I hear about him before? I lifted my shoulders. "You have another brother? Who is he?"

Dad stared at the picture with a blank expression as he said, "His name was Eugene. He died a long time ago." I absorbed this information. My uncle Eugene was dead, so that's why I didn't know him. But I reasoned he must have had children before he died. We visited aunts, uncles, and cousins, so I assumed I would know this new uncle's kids. If not, I wanted to meet them.

I touched Dad's arm. "Who're his kids?"

Then came the bombshell. Dad put my hands together and surrounded them with his. "He didn't have any kids. He was only four when he died."

I gasped and stepped back. "Four! Littler than me?"

Dad furrowed his brow and explained, "Eugene died because a heavy mirror fell on him."

Tears streamed down my cheeks. I couldn't understand how my father was so calm. "Why aren't you crying?"

Dad raised an eyebrow and steepled his fingers. "I'm not crying because he died a long time ago. And anyway, I didn't even know him."

I wrung my hands. "But everyone knows their brother."

Dad took me in his arms and patted me as he clarified, "I didn't know this brother because he died before I was born."

This I could understand. If Eugene died before Daddy was born, then of course he wouldn't miss him. But what about the pain his parents must have felt in losing a young child?

I only knew of these grandparents, David and Marion, from Dad's stories, because they died long ago. According to Dad, his mother had a knack for making the ordinary extraordinary. Her meals were artistic statements: she coiled carrots, sculpted radishes, swirled mashed potatoes, added dabs of ketchup or mustard for balance and interest, and completed her culinary presentations with parsley or other greens. She decorated every room in her house with tasteful art and classic furniture.

Marion was also a gifted seamstress who fitted and designed dresses from their home. One essential tool was her floor-standing mirror.

CHAPTER ONE

Marion and David with Eugene. FAMILY PHOTO

On the morning of August 10, 1916, Marion was sewing as four-year-old Eugene played on the floor. Marion was stitching white lace trim on a dark blue dress and wanted it to be perfect. Eugene practiced somersaults: head up, head down. His straight hair flapped, and he laughed when he completed each tumble. He was good at flips and almost never lost his balance. But that day his equilibrium failed him. His legs came tumbling down and struck the massive floor-standing mirror. The mirror teetered back and forth until it crashed down on the little boy. Eugene lay motionless. In the time it took Marion to cross the room her son was dead.

When Dad told me about this terrible accident, I already feared death because my mother was recovering from a severe illness. Before she came home from the hospital, I overheard neighbors in our Brooklyn apartment building whisper that she might die. Even after she came home, I still worried about her. I clung to the belief that she wouldn't die because children seldom lost parents. But I also knew it was rare for a child to die. Hearing about my uncle's death when he was only four introduced exceptions to these norms, and this frightened me.

My mother had become ill the previous summer when I was five. I learned about it while at a backyard picnic with my older sister's friend's family. I didn't know them, and my parents weren't there. The father grilled hamburgers and as the mosquitoes attacked me, the mother put a huge serving of potato salad on my plate. "Here you go, Susan, enjoy!"

"Thank you," I said, trying not to throw up. My parents knew I hated potato salad. I didn't like hamburgers either. I drowned the burger in ketchup to make it edible, nibbled at it, and moved the potato salad around on my plate with my fork.

I looked at the strangers serving me dinner and slumped my shoulders. "Where are my Mommy and Daddy? When are they coming?"

"They should be here soon," the friend's mom said while sipping her Coke. "We have a special surprise for you later."

I scratched the red welts on my arm and wished I were home. Ten-year-old Diane and her friend whispered and laughed at the other end of the table. The mosquitoes didn't seem to be bothering them.

We were in Free Acres, New Jersey. We spent summers in this country town staying with my mother's cousin to escape Brooklyn's city heat. Dad joined us on weekends but stayed in the city during the week. My sister and I attended the local Wanoga Day Camp. Although most of the other campers lived in Free Acres year-round, Diane fit right in. She was a natural athlete, a talented dancer, and outgoing. One year they cast her as the Pied Piper in a silent movie the entire camp put on and filmed. I was proud to be her sister.

CHAPTER ONE

Making friends at Free Acres was harder for me. Most of the kids my age were boys who loved sports. There was only one girl, and unlike me, she was athletic. Still, she was kind and was my friend. In camp, I learned to swim, play ball, and enjoy nature. I especially loved the camp's arts and crafts, which kindled my lifelong interest in art.

Mom and Dad appeared at the end of dinner. I moved over to make space for them on the picnic bench, but only Mom sat down. She looked exhausted. Her arms and legs were fat even though she was skinny. And the skin around her eyes drooped. Even though Mom looked terrible, seeing her there gave me hope.

Dad tightened one hand into a fist and cupped it with the other hand. "We can't stay for dinner. We just came from the doctor." Dad furrowed his brow and looked back and forth at Diane and me. "Mommy is sick. She needs to go to the hospital."

"What's wrong?" I bunched up my napkin and put it on the table. All my hope evaporated. "Why can't she get well at home?"

Dad took a deep breath with a hint of a sob. "Water is swelling Mommy's arms and legs because her kidneys aren't working right."

Why was Daddy talking in riddles? I rubbed my itchy bites. "What are kidneys? Why can't the doctor come to us? When I was sick, he came to our house."

Dad rested his hands on Mom's shoulders and squeezed her. "The kidneys remove extra water from the body." He smiled wanly. "They can take better care of Mommy in the hospital."

This didn't sound right. Mommy should be at home with us. I scratched my leg and looked at my sister Diane for support. Her mouth and eyes were open in shock, but she remained silent. I twisted my short ponytail. "How long will Mommy be there?"

Dad answered my questions, letting my mother just listen. He cleared his throat. "She can't come home until she gets well. Nobody knows how long that will take."

What? He can't even tell me when she's coming home. I looked at my mother for help, but she cast her eyes down.

Dad turned to the other parents. "Thanks so much for watching the girls for us. I'll come back for them as soon as I get Cyrilla admitted."

Diane's friend's father stepped forward and clutched his hands together. "Don't worry. We'll drop your daughters off after we take them to play miniature golf."

My parents started to leave as the friend's mother began serving dessert with a bright smile. "What flavor popsicle do you want? Hurry. We have reservations to play miniature golf in half an hour."

I didn't understand why she sounded so happy and why she talked about miniature golf. This was so wrong. Didn't she hear what my daddy said? Didn't she know my mother was very sick? It wasn't right that no one seemed upset about this news. Won't Mommy feel bad if her illness doesn't make us sad? I decided that I needed to cry. Doing my best to look sincere, I sobbed and called after my parents, "I don't want Mommy to go to the hospital! Please, can't she stay home with us?"

My sister and I lovingly called our parents Mommy and Daddy when we spoke of them, even as adults. I didn't see Mommy again for months. She was gravely ill with nephritis—a disease that causes inflammation of the kidneys—and was in intensive care at a hospital in Manhattan. She remained there when we returned to Brooklyn, and I entered first grade.

Daddy visited her every night. The doctors told him they didn't expect her to live. My father dressed Diane to look older so she could see Mom, but they didn't allow me in because I was too young. Instead, Dad brought my mother the arts and crafts items I made for her.

When it seemed my mother would never get well, the specialists tried cortisone, an experimental drug at the time. At first the dose was too high. Dad was in Mom's room late at night and could see from the monitors that she was fading. He panicked and insisted the nurses wake her doctor who then rushed to the hospital to lower the drug dose. This miraculously led to Mom's recovery. From then on, she was on a strict low-sodium diet with many special foods, including a milk substitute and bread made without salt, both of which tasted awful. Even after she returned to work, Mom rested in bed each afternoon when she came home, and we continued to treat her like an invalid.

In 2022, Russia invaded Ukraine. My Ukrainian ancestors' foresight, struggle, sacrifice, and courage saved me from this carnage. They had left everything familiar and started over in a new country with a new language.

My paternal grandfather, David, was nineteen when he immigrated to Scranton, Pennsylvania, in 1906 with his parents and two older brothers. They came from Kharkov, the second largest city in Ukraine, and often described its wide streets and enormous squares.

At various times, Kharkov was in the Russian empire, Poland, or Russia. When my family lived in Kharkov it was in "Little Russia." They spoke Russian and listed Russia as their country of origin on official papers.

Jews were supposed to live in the Ukrainian Pale of Settlement. But many Jews, including soldiers, students, artisans, tailors, and merchants, had special permission to live outside of the Settlement, in Kharkov. Some of these Jews held important positions in banks, medicine, engineering, and the arts. My ancestors were tailors with only a few years of elementary school education. The population was about 35 percent Ukrainian, 55 percent Russian, and 10 percent Jewish.

In the 1880s and from 1903 to 1906, Jewish men, women, and children in the Ukrainian Pale of Settlement were butchered in antisemitic pogroms. Jewish babies were torn to pieces before their parents' eyes. These vicious murders of their nearby brethren terrified the Kharkov Jews. Although they were spared from these atrocities, they were still subject to severe antisemitism. In 1893, local guilds forced sixty-seven Jewish craftsmen and their families to leave Kharkov, followed by other expulsions of Jews who held important positions. The Black Hundreds antisemitic movement (1906–1914) founded in Kharkov denied Jews even the harsh choice of baptism over death, an option that was previously available to them. They simply murdered Jews.

Countless Ukrainian and Russian Jews, including my family, left Ukraine and never looked back. They and their children refused to be victims. Instead, they embraced opportunity and started anew.

My grandfather's brothers were skilled tailors. One started a successful tailor shop to make custom suits for Scranton's most discerning customers. The other worked with fur. According to my uncle Cyrus, the furrier was a little temperamental, but his talent was in such demand that eager customers competed for his attention by driving him home for lunch each day, at first with a team of two horses and later in a Reo motor car.

Unlike his older brothers, my grandfather, David Weissman, was not ready to settle down and open a business when he first arrived in Scranton. He shortened his name to Weiss, then left home to seek adventure. He decided to make his fortune on the railroads, which were expanding throughout the United States. David explored Carbondale, Pennsylvania, where crews loaded train cars with coal for transport to New York City. There he became close friends with the grandson of a Civil War hero who used his connections to help David get a job on the Delaware-Hudson Railroad. This railroad made history when it carried the first paying passenger in North America.

My grandfather married Marion Millner when he was twenty-five and she was nineteen. The marriage was a happy one. Marion had shiny chestnut-colored hair, soft flush lips, and large brown smiling eyes. Still, her elongated oval face, pointy chin, and large nose gave her a homely appearance. In contrast, her husband was movie-star handsome with brown hair, straight nose, firm jaw, and gray-green eyes. He also had a splendid physique. He was a short man, yet taller than his wife.

They moved in with David's friend and his wife who lived in a large house on Grove Street. Eugene and my father Norman were both born in that house, which was a fifteen-minute walk from downtown Carbondale. Later, the family moved into their own place in the center of the city. Soon after Eugene's accident, my grandparents left Carbondale. My grandfather gave up his dream job with the railroad, and they restarted their lives near his brothers in Scranton. I learned much later that a secret associated with Eugene's accident necessitated this move.

Three years after their return to Scranton, Grandma Marion's mother, Sarah, passed away at the age of forty-seven. She left an eight-year-old daughter, Sylvia, who joined her older sister Marion's household, and my uncle Cyrus was born five years later, completing the family. Eugene's accident and Sarah's early demise caused my grandparents to worry about sudden death.

After these tragedies, the family curse wasn't obvious for a century. However, looking back there were illnesses and deaths that could have been caused by a deadly mutation. Since many of my relatives died young or were seriously ill from many causes, it would be years before I could use my expertise as a geneticist to tease out the trail of the deadly mutation. In time, science and technology equipped us with simple, inexpensive genetic tests that we were able to use to complete the story and save lives. Such tests can now discover disease-causing mutations regardless of ethnicity or family health history.

TWO

Catastrophe

> "Your goal each day should be to be satisfied. If you don't have a catastrophe, you should be grateful." —My father, Norman Harold Weiss

If a deadly mutation caused my niece's death, it could have come from either my mother's or my father's side of the family. Looking at their history, either was possible.

It was the spring of 1929, thirteen years after little Eugene's accident and the family's move to Scranton, Pennsylvania. By then my father was fifteen, and he and Grandpa David went to a father-and-son banquet sponsored by the Striver's Club. The Young Men's Hebrew Association (YMHA) was to award Dad the Boginsky Medal for character and service at the banquet. Grandpa wore a tuxedo. My father donned his first suit. He was taller than his father. Their dress shoes clicked on the sidewalk with each step as they ambled past cherry blossoms. When they arrived at the hall, flowers adorned the tables. During dinner, officials praised Dad in speeches and called him up to receive his medal.

CHAPTER TWO

Because of Dad's personality and outstanding record in service, sports, and academics, his classmates voted him the most likely to succeed. Dad had Harvard University in his sights, despite Harvard's bias against accepting Jews. His parents could afford the tuition because the Lackawanna Upholstery Company my grandfather founded after moving away from his job on the railroad in Carbondale was a financial success. Also, my grandmother's dress business prospered.

Five months later, the stock market crashed. The Great Depression upended lives throughout the country. Businesses closed, putting millions out of work. Men sold apples and pencils on street corners. Some jumped out of windows. By 1930, few people could afford to buy or repair furniture, and David's upholstery business was no longer profitable. My grandparents lost their savings when their bank failed, and they could no longer afford to send Dad to an out-of-town college. With no remaining business ties to Scranton, the family moved to be near my grandmother's brothers in Brooklyn and posted a dressmaker sign in their apartment window to attract customers.

Tragically, on August 14, 1930, Grandpa David unexpectedly passed away at forty-two. Dad became the man of the house and surrogate father for his six-year-old brother, Cyrus.

Here is my uncle Cyrus's email recollection of a day spent with my dad, Norman, five years after their father's death, when Cyrus was eleven and Dad twenty-one.

> As I approach ninety years of life on this mortal coil, I think of many memorable events, like the day in Brooklyn long ago, when the sky was very blue and the grass very green. My brother Norman, in the depths of the Depression, told me to get dressed. We were going to the baseball game. Later that day, I watched as he produced a miracle, showing the man at the ticket box a dollar ten for two bleacher seats at Ebbets Field. Norman Harry Weiss, the Erasmus Hall [High School] athlete, my brother, my acting father, my mentor, my hero, was not finished. He then comes up with two nickels for peanuts and Coke.

Dad, his younger brother Cyrus, and father David. Taken in front of their apartment on Eastern Parkway in Brooklyn in 1930. The sign in the leftmost downstairs window says "WEISS DRESS SHOP." FAMILY PHOTO

CHAPTER TWO

The stadium was crowded, as this was a special day. We bums were hosting the Boston Braves who had just acquired the Sultan [Babe Ruth], the Great One, the Bambino himself. He was to play out his final days for the Braves. Our bleacher seats were so near his position, I felt I could touch him when he trotted out to the outfield, doffing his cap to the thunderous cheers. He was looking directly at me. I grasped Norman's hand as tightly as I could, and for several moments I enjoyed a happiness, a remembrance of which will forever be with me. I do not know who won the game that day, nor can I remember what the score was. However, after the game, on the walk up the Franklin Avenue hill, I looked up at Norman still, clutching his hand, and knew no harm could come to me as long as he was near. How lucky I was to have had him.

Reading this made the impetus for Dad's teaching—"Your goal each day should be to be satisfied. If you get through a day with no catastrophe, you should be grateful"—clear to me. After losing his father so young and suddenly, Dad learned to celebrate every ordinary day devoid of tragedy as a spectacular gift. I was reminded of this when A. A. Milne's Eeyore expressed appreciation in *Winnie the Pooh* that there weren't any earthquakes that day.

If there were a deadly mutation in the family, it could have come from my mother, and she could have gotten it from her mother, Jennie, or her father, Simon. I know about Simon's mother Elke through family stories and the exquisite items she left to us. Among the things she carried from Czarist Russia when she immigrated to Brooklyn was a pair of cherished Shabbat candlesticks that are now a centerpiece in my home. Their etched myrtle flowers illustrate love of life, and their engraved grapevines suggest the passing of this love from generation to generation. When I lift the massive-looking heirlooms to prepare for Shabbat, I'm shocked to find they're lightweight and hollow. Their thin sterling silver encasement parallels the spiritual strength that shaped Elke when tragedy left her empty.

Victor and Elke married in 1870 in Czarist Russia. The arranged match between two teenage cousins proved disastrous. Divergent views on the meaning and purpose of life divided them. Elke was a deeply religious daughter of the renowned Rabbi Jacob Simon. Victor rejected traditional religion with its laws and customs practiced by his orthodox family. Instead, he put his energy into political activism. He was a revolutionary.

Although Alexander II freed the serfs in 1861, they still lived under extreme hardship in 1870 with a heavy tax burden. Victor organized the peasants and encouraged them not to pay taxes. In his fervor, he ignored family responsibilities, creating a constant unbearable rift between husband and wife.

The marriage produced my grandfather, Simon, and his sister (whose name is unknown) before ending in divorce. Victor took Simon, and Elke raised the girl. Victor's revolutionary activities kept him too busy to care for his son, so he gave Simon to his parents, who showered him with love and attention. Since Russian schools forbade Jewish enrollment, Simon's grandfather gave him a solid education at home.

Still young and attractive, Elke remarried. Once she established a stable home, she wanted her son Simon to come and live with her. But he didn't want to leave his grandparents. Elke bore five more children with her second husband.

Victor had another short-lived marriage that resulted in a son, Noah Hyman. After their divorce, his second wife and child immigrated to the United States with hopes of a fresh start. Unfortunately, the mother was confined to a mental institution. Even so, Victor remained unwilling to compromise his political activities to care for Noah. Since his parents raised Simon for him, he expected them to do the same for his second son. Older now, they refused, so Victor put Noah in an orphanage.

Victor practiced "Tikkun Olam" translated as "repair the world." Scholars interpret some Jewish texts as asking us to partner with God to complete and repair his creation through righteousness, justice, and compassion. However, Victor didn't apply those principles to his own family.

Once Simon's grandfather taught him everything he knew, he encouraged Simon to go to the United States to continue his education. Since they didn't have enough money for a ticket to take him across the Atlantic, his grandfather reminded Simon, "You know you have a mother too. You should ask her to help you get to America."

With Elke's many children, I don't know how she financed Simon's trip. But she did. Simon continued his education and became a dentist in the United States. He displayed his framed New York City license number 711 to practice dentistry for his patients to see.

Catastrophe struck Elke's new family when she was about forty. There was a diphtheria epidemic. Her husband and all six children living with her died that year. Most succumbed within a month.

Despite her grief, Elke refused to give up on life. Her son Simon invited her to immigrate and live with him in the United States if she would adopt her despised first husband's name, Zuchtmann, which would give mother and son the same last name for propriety's sake. Elke agreed and summoned the courage to embrace a new future.

When she came to America in 1896, she brought three valuables among her meager possessions: Victorian silver and brass Shabbat candlesticks as well as an engraved gold watch, and a pendant set on a dull eighteen-karat chain. The family still treasures these.

Elke lived with her son Simon and kept house for him until he married at thirty-nine. Simon's bride, Jennie Hymanson, is the only grandparent I knew. Before she got married, she taught in public and Hebrew schools. By the time she wed Simon Zuchtmann, a dentist fifteen years her senior, she was twenty-four and almost an old maid.

After Simon married, his mother Elke moved in with her sister to allow the newlyweds privacy. Elke's sister was an invalid, so Elke helped raise her niece and nephew. Elke still lived near Simon and lavished attention on his children. My mother and her younger brother Herb well remembered and adored their Grandma Elke. She seemed young and vital to her grandchildren with her large sensitive eyes, small straight nose, and flawless fair complexion. She wore a brown sheitel wig to hide her hair from men outside the family, following the custom for married Jewish women. When Herb saw her once without her wig, he was frightened because he didn't recognize the old lady with long white hair.

I often think of Elke. When I use her candlesticks to *bench licht* (light Shabbat candles), I am transported to what I imagine of her shtetl life. There, at dusk on Friday nights, amid savory aromas, surrounded by her husband and six children, Elke lights candles, head covered, eyes closed, hands undulating over flames. Elke's chant of the ancient Bracha prayer beckons us to her island in time, Shabbat peace.

Elke served as a role model for future generations in our family. Her story and example helped those that followed deal with the adversity and losses that awaited them. She taught us to "choose life—so that you and your children after you will live" (Deuteronomy 30:19).

THREE

Interlude: Genetic Testing of Infants

"Blessed will be the fruit of your womb." —Deuteronomy 28:4

The reason I wanted to find my family's deadly mutation was to extend and improve the lives of those who inherited it. The goal was never to warn anyone not to have children. As a society we "choose life" by enacting rules to protect our children's health and welfare. Doctors now screen babies in utero and as newborns for various genetic disorders to warn parents of severe conditions and allow for intervention before the onset of certain diseases. When my sister and I were born, there was no such screening. If extensive newborn genetic screening and current medical treatments had been available, it would have had a dramatic effect on us. My parents would have learned that my sister had a potentially lethal (although usually adult onset) mutation. Unless this was a new mutation in my sister, she would have inherited it from one of her parents. Because of that, doctors would have recommended that my mother and father undergo testing themselves, and they would have found that one of them carried the mutation. These tests would have warned my sister and the affected parent to get preventive treatment to improve and lengthen their lives. Also, my niece, Karen, would have known she had the mutation and would have been treated and might still be alive.

On the other hand, knowledge of the mutation could have affected our lives negatively. In the extreme, fear of sudden death could have caused my parents to not have a second child. In that case I would never have been born. My sister's fear of the increased risk of early death might have caused her to avoid marriage and children. If she did have children, her daughter Karen's knowledge that she carried the mutation could have caused her to be less adventurous.

Genetic counseling could have helped us understand the risks. The counselor would have explained that having the mutation did not guarantee that the patient would actually get the disease. Sometimes even if you have a disease-causing mutation, you never exhibit any disease symptoms. The chance of a mutation causing any effect is its penetrance. The mutation in my family has a high penetrance; it affects the heart in over 90 percent of patients with the mutation. However, although most patients have some effect, the problem the mutation causes is often relatively minor, such as an arrhythmia only detected by medical tests. Indeed, our mutation causes the extreme of sudden death in only 15 percent of the people who carry it. Thus, we should view genetic information as one aspect of our medical history that can help us take precautionary measures to prevent disease or postpone its onset. Genetic results can also bring relief to those with a family history of disease if they learn they didn't inherit the mutation and are therefore not at risk. There is currently interest in extending these screens to include many more genes so more children will live.

Expectant parents or couples who plan to have children often undergo reproductive carrier screening for mutations that could cause a disease in their babies. We all have two copies of

INTERLUDE: GENETIC TESTING OF INFANTS

every gene. In carriers, one copy of a gene works properly while the other copy has a mutation preventing it from doing its job. If the mutation is recessive, the good gene copy keeps the carrier healthy. Reproductive screening warns couples if they both have recessive mutations in the same disease gene. This would put their baby at a 25 percent risk of inheriting a bad copy from both parents, resulting in a genetic disease such as cystic fibrosis, sickle cell anemia, or Tay Sachs. Some parents facing this risk choose adoption; others use a surrogate mother or donor sperm. Parents can also choose in vitro fertilization and pick a healthy embryo to implant into the mother's uterus. Doctors understand and promote reproductive carrier screening.

There are many efforts to screen and even eliminate recessive genetic diseases. Dor Yeshorim, also called the Committee for Prevention of Jewish Genetic Diseases, is a program that tests people, identified only by number, for a variety of recessive gene mutations common among those of Ashkenazi Jewish descent. They then warn couples who plan to start courting to seek other partners if they have recessive disease mutations in the same gene. This approach has effectively eliminated Tay Sachs from the Orthodox Jewish community.

However, reproductive screening doesn't report on dominant mutations that cause problems even in the presence of a good copy of the gene. Such mutations also put the parent at elevated risk for disease. I asked several organizations that provide reproductive screening why they don't screen for dominant mutations at the same time. They explained that some parents would refuse testing altogether if they had a chance of learning that they themselves were at genetic risk.

The first widely used test to screen infants for a genetic disorder detects an excess of phenylalanine (an amino acid that is found in most proteins) in a drop of blood. This warns of phenylketonuria (PKU), a condition that causes brain damage without treatment. Dr. Robert Guthrie invented this test in 1962.[1]

Putting infants found to have PKU on a diet low in phenylalanine prevents mental impairment. This dramatic success led to the screening of newborns for other diseases.

The Department of Health and Human Services issues a list of illnesses recommended for screening called the Recommended Uniform Screening Panel,[2] although each state has its own specific newborn screening panel. For a condition to be on the list, there must be a test that can detect the condition within one to two days after birth. Also, to be on the list, early detection must benefit the patient. In 2024, each state tests its newborns for between thirty and sixty rare but serious health conditions from that list.

The advent of rapid and inexpensive DNA sequencing may soon allow us to screen for thousands of disorders babies are at risk for during childhood. Since there are so many rare genetic diseases, it is fairly common (3–6 percent) for a baby to have one. Currently, there are treatments for only about 10 percent of the six thousand known genetic diseases.[3] The number of treatable diseases is expected to increase dramatically as new technology that allows us to correct mutations enters the clinic.

Even in the absence of available treatment, there is value in identifying mutations that cause disease. This allows patients and their families to build community and together look for treatment. This was the case for Lilly Grossman, who suffers from a movement disorder. When genome sequencing identified a mutation in her *ADCY5* gene, they put her on Diamox because that drug helped another patient with an *ADCY5* mutation.[4]

The idea of using genomic sequencing in the routine care of newborns is gaining momentum.[5] The cost of DNA sequencing continues to fall, and models of economic implications show that genome screening could be cost-effective by preventing expensive diseases. Over 10 percent of babies sequenced are at increased risk for genetic disease and about 2–4 percent already show symptoms.

Clearly, sequencing a newborn's DNA is fraught with ethical challenges. Should a parent have access to their child's genetic information without the child's consent? Should treatable genetic conditions that do not show up until adulthood be part of a newborn genomic screen? This would not only protect the infant later in life but could also alert the parents who would be at risk for the condition. Should illnesses without treatments be part of newborn DNA sequencing? Should we make sequencing available to all newborns, paid for by the government? Should it be compulsory? The costs of follow-up clinical care must also be affordable to all. Insurance should be required to cover this since preventive care of this type is generally cost-effective compared to treatment after illness appears.

Another concern is the importance of keeping genetic information private to avoid social stigma, job discrimination, or difficulties with insurance. Except in rare circumstances, it is illegal to consider genetic results when issuing health insurance in the United States. However, there is no such protection for the purchase of life, disability, or long-term-care insurance. We need legislation that allows people to buy the insurance they need without allowing them to game the system by buying excess insurance if they learn from a genetic test that they have an increased likelihood of needing the insurance.

My family's experience kindled my interest in learning how to recognize and treat genetic disease. Despite the ethical hurdles associated with sequencing newborns' genomes,[6] it is essential that we keep our eyes on the prize. We must find a way to use genetic tests and screens to help patients.

FOUR

Family Life

> "Of all the means to insure happiness throughout the whole life, by far the most important is the acquisition of friends."
> —Epicurus

According to Harvard's eighty-year study of happiness, strong relationships are more important for happiness than anything else, including genes.[1] My family counterbalanced the harm from our genes with strong relationships. The story of our parents' acquaintanceship at Erasmus Hall High School fascinated my sister and me. Before we enrolled in that same Brooklyn school, we wondered if we would meet our husbands there too.

Among the three hundred pupils in their 1932 class, Dad was the brain and Mom the beauty. They only met because their teachers assigned seats in alphabetical order. There weren't enough chairs in the classrooms for all the students. My father, Norman Weiss, was often the last to get a seat and my mother, Cyrilla Zuchtman, rarely got one. She sat on the floor or took notes standing and leaning her notebook against the wall. Dad

CHAPTER FOUR

used this opportunity to spark friendship at the end of the alphabet by offering Mom his seat.

After graduation, my parents attended City College of New York (CCNY) night classes, commuting together on the subway for an hour in each direction. Many of the smartest students attended CCNY during the Great Depression because they couldn't afford out-of-town tuition. Competition for admission was fierce.

When Cyrilla first met Norman she was at once aware of the pimples covering his cheeks and forehead. But, by the time they became casual friends in high school and later confidants traveling to college together, she no longer noticed the acne. Her admiration for his accomplishments, ambition, and concern for others unconsciously guided her into only seeing the twinkle in his eyes and his friendly smile.

Over the months, as they balanced themselves in the shaking subway car on the way to CCNY, Cyrilla learned Norman went to college at night because he worked days at his uncles' print shop, The Millner Brothers. Over the loud rumble of the train, Norman told Cyrilla he started as an errand boy and progressed to become the office manager. His experience with printing presses, typesetting, and business practices later enabled him to open his own print shop.

Cyrilla attended CCNY at night because her grades weren't good enough for her to land a spot in the day session. Her academic troubles were unexpected—she was smart and conscientious. Decades later, she realized her problem was dyslexia. During her youth, doctors and educators knew little about diagnosing and treating this disorder. Even though she struggled in school and dropped out of college, Norman respected her intellect. He discussed current events, religion, and books with her, boosting her self-confidence. He also made her feel at ease with his silly spontaneous puns. Her complaint about filling out a long form would get a rejoinder like, "Didn't your father the dentist teach you the drill about fillings?"

Norman had to work to help his dressmaker mother, Marion, support his little brother Cyrus and his aunt Sylvia, who lived with them, because his father was dead. Cyrilla recognized his responsibility for Cyrus. She also had a much younger brother she cared about. But she didn't understand why Sylvia, who was three years older than Norman, didn't help support the family. Marion forbade this, thinking it improper for a single girl to work. Mom stared in disbelief with a clenched jaw when she heard that. She didn't think it was at all fair that Sylvia wasn't sharing Norman's burden. Single men also needed a life outside of work and school. Cyrilla took this as evidence that Marion favored Sylvia over Norman.

Cyrilla told Norman that favoritism made her especially angry because her mother gave her brother preferential treatment. She also complained that her mother jeered at her dentist father, Simon, whom she idolized. Because people seldom saw dentists during the Great Depression, Grandma Jennie taunted her husband, in a sing-song voice, "What good did it do for me to marry a professional man?"

Cyrilla often wrinkled her nose at cigarette smoke when they waited for the train at Prospect Park station. She fiddled with the watch hanging around her neck that she inherited from her grandmother, Elke, and gushed about picnics in the park and boat rides on the lake with her different suitors. Cyrilla had delicate health, so her dates did the rowing to save her energy and protect her hands from the rough oars.

After boarding the crowded subway car, Norman and Cyrilla would hang on to straps as the train sped past the DeKalb, Grand, and Broadway stations. He'd look upon her captivating smile, jet black curly hair, and flawless complexion as they continued their conversation. They regularly found

CHAPTER FOUR

seats when the subway doors opened at Washington Square and a flood of riders disembarked. As she sat, Cyrilla would tuck in her flare skirt and swing her small feet, clad in medium-high heels, under the bench. Above the roar of the track, she talked about other dates to the movies or dances. Norman took mental notes about what she did and didn't like. She praised intelligence, kindness, and athletics. Indeed, Norman's success as a track team star and his racing in the famous Millrose Games at the Armory impressed her.

For several months, Cyrilla spoke daily about one admirer in particular. This young man was handsome, had an English accent, and carried an umbrella. Norman was jealous and worried he would lose Cyrilla's heart to this Englishman. One day as they rode past the Columbus Circle station, she confided with tears in her eyes that this suitor had stopped calling on her. She said, "My father asked him, 'Young man, what are your intentions with my daughter?' and I never saw him again after that."

Norman picked his chin up and threw his shoulders back as he seized this chance to turn their friendship into something more. "Why don't you give *me* a try?"

Her eyes widened. She thought this was a joke. How could she date Norman? He was such a brain and was just a friend. But he persisted. He knew he had qualities she admired from listening to her stories about her dates. Before long, she relented, realizing she was already in love.

When they married, my mother was twenty-one and my father twenty-two. Mom hired a photographer for the wedding even though they couldn't afford one. She planned to keep the proofs and not order or pay for finished photographs. Indeed, a framed 8" X 10" sepia bridal portrait with small holes punched across the picture stating "PROOF RETURN TO . . ." sat on my parents' dresser for decades.

When I asked about this, Mom put her hands in her pockets and said, "The photography store went out of business, leaving us the proofs." I learned the truth from Dad. Mom licked her lips as she defended herself, saying, "Everyone did it. No one could afford a photographer during the Great Depression."

With a modest smile, Dad said, "I didn't approve. That's why I'm not in any of the pictures." My mother respected Dad's honesty. In return, Dad didn't judge Mom, which allowed him to enjoy her bridal portrait. They never tried to change one another. Both had the same goal of having happy children who succeeded in life. They were willing to work hard and sacrifice their own comforts to achieve this. Mom's stories about her boyfriends showed Dad she understood people and valued kindness, education, and ambition. He loved that she shared her thoughts and experiences. This helped him reciprocate and reveal things about himself. She was the perfect loyal confidant. Dad enjoyed being her protector.

Mom yearned to teach at a public school, but she found college too difficult to achieve this dream. Instead, she became a school clerk and steered Dad to take the education courses required of teachers. She thought his uncles were taking advantage of him at their print shop and that he should instead become a teacher. She narrowed her eyes. "Teachers in the public school get tenure. Who knows how long this depression will last? A teaching job is very secure. And you could become a principal."

After seven and a half years at CCNY night school, Dad graduated summa cum laude with his master's degree in 1940. He easily passed the teaching exam, but his Pennsylvania accent prevented the school board from hiring him. They claimed pupils would have difficulty understanding him, so Mom helped him shed the accent.

CHAPTER FOUR

He joined the faculty at Central Commercial High School, where he taught bookkeeping, keypunch, and punched card tabulating and became a department chair. Mom was a clerk in an elementary school but continued to admire and envy teachers.

With Dad's administrative supplement they could, at last, afford a baby, but cycle after cycle Mom's failure to conceive disheartened them. After trying for a year, they consulted doctors, who told my father to wear loose-fitting boxer briefs to raise his sperm count.

Two months after the Pearl Harbor attack, Mom was with child. By then, Dad volunteered for the Sector H Civilian Defense Unit, where he designed and taught courses in office machine operation.

Mom wanted to protect Dad from the draft by having a cesarean section to give the baby an earlier birth date. She correctly thought the Selective Service would only allow fathers who conceived babies before we declared war to be exempt from the draft. My father wouldn't allow this operation because of the risk to Mom and the baby, so my sister Diane was born following a normal pregnancy. Dad drew an early draft number in the lottery and the Army inducted him when Diane was less than a year old.

This devastated Mom. She opposed all wars, no matter the cause. In her view, politicians used war to gain power, and soldiers were their pawns. She was terrified that Dad's being in the Army would put his life in danger, and she was heartbroken it would separate him from her and baby Diane.

Mom expected her mother-in-law, Marion, to understand and sympathize. Instead, Marion declared she was proud her son was a soldier and told Mom she should stop feeling sorry for herself. Marion had lived through worse.

Dad's extensive background in office machines and top score on the Army placement exam landed him a position in the 44th Mobile Machine Records Unit in George S. Patton's Third Army. There were forty-one men from all over the country in this elite unit. Dad was the only city boy. A newspaper writer in the group described him as a "driving whip of energy from the sidewalks of New York."

After Dad's unit left for Europe, Mom and my sister Diane, who was still a baby, moved in with Mom's parents, Grandpa Simon and Grandma Jennie. It was very hard on my mother when her father, Grandpa Simon, died of cancer a few years later while the War still raged. Mom's kid brother, Herb, whom she affectionately called Herbie, lived there too and helped care for Diane.

One day when Uncle Herbie sat in the living room with his feet up on an ottoman, Diane came home from nursery school and showed him a baby chick the school bus driver gave her for Easter. Diane hopped from foot to foot as she showed Herbie her new pet. This live animal was much cuter than her stuffed toy creatures. She played with rubber ducks in the bathtub, so she decided a bath would be good for her real chick. She leaned over to reach the faucets, filled the tub with cool water, and put the chick in. When the bath ended, she took him out and dried him off. She wrinkled her nose as she noticed the chick wasn't chirping or moving anymore. It's strange the chick fell asleep in his bath, she mused. "Look, Uncle Herbie, the chick's sleeping!"

But Uncle Herb knew better. He gathered Diane into his lap and explained that the chick wasn't sleeping, but that she had drowned it in the tub. Diane's eyes widened and she began to bawl. This rude introduction to the fleeting nature of life was

Picture of Mom and Diane that Dad carried during the War. FAMILY PHOTO

FAMILY LIFE

traumatic. Diane never forgot it. Throughout her teenage and adult years, she worried that every bump on her body was cancer and every minor illness would prove fatal. I think the trauma with the chick and all of Diane's false illness scares were rehearsals that gave her the strength to survive the genuine illnesses and deaths in her future.

As far back as I can remember, my sister and I loved to secretly read the War letters that Mom saved in a box on the top shelf in her closet. As I touch the originals yellowed with age, I feel my parents' history. I recognize Dad's large, round handwriting confirming that he was the author. His V-mail (short for Victory mail) is on a sturdier stock. The Army photographed and transferred V-mail to microfiche to save space during transport overseas. Before delivery, they printed a legible but size-reduced version of the original on cardstock along with a censorship stamp.

The 44th boarded the USS *Billy Mitchell* in New York City on May 2, 1944, and the ship arrived in Scotland two weeks later. Fifty days after D-Day they landed in Utah Beach, Normandy, France. For security reasons, the return address on Dad's V-mail states, "Somewhere in France." V-mail dated after March 1945 says, "Somewhere in Germany." The Army soon promoted my father to sergeant. Although Dad missed my mother and sister, he otherwise adjusted to Army life. Living through the Great Depression had prepared him for hardship. Although the unit traveled with the front-line soldiers and the Army trained and armed the 44th for a firefight, my father was thankful he never faced combat.

He tried to be upbeat in his letters, but sometimes he shared his sadness and loneliness. "How long I'll be stuck in this damn Army–God only knows. Actually, it's not too bad with a unit such as this if only I could get to see you even once every two months." The letters always ended with, "Kiss the love girl Diane for me. Love and a million scillion kisses."

My father also corresponded with his friend Marvin Fein during the War. Before they entered the Army, the two men were department chairs at Central Commercial High School in Manhattan. Through their letters, they hatched the idea of founding a printing business together when the War was over. Planning for a return to normal life helped them see beyond the nightmare they were living as soldiers.

Dad was well respected for his leadership and innovation as a soldier. The Army awarded him a Bronze Star for this. Also, toward the end of the War in Europe, the Army offered Dad a field commission. He welcomed the extra responsibility and pay but feared accepting the position could delay his return home if they sent officers to Japan following Germany's defeat. He wrote to Mom for guidance but had to give the Army his decision before getting her reply. Not wanting to risk prolonging time away from his family, he turned down the commission. Also, he wanted to be a regular soldier and eat with his men rather than with officers.

He wrote, "While we live in barracks and eat in a mess hall, the officers live in a hotel, have room service, tablecloths, and waiters. Don't think I envy them, because I detest the class distinctions in the Army and would be unhappy if I were any party to it."

Dad had the same response to class divisions among Jews. Years later I learned in Hebrew school that ancestry separates Jews into three groups: Kohanim, Leviim, and Yisraelim. The priests' descendants are Kohanim, and sanctuary workers' descendants are Leviim. Jewish prayer services afford these two groups special honors. The Yisraelim are everyone else.

I asked Dad, "Which group does our family belong to?" I expected us to be in the most exclusive Kohanim group. It disappointed me to learn we're "just" Yisraelim. This reaction surprised Dad. "But, Susie dear," he said, "Yisraelim are the regular people and are the best group to be in." Since then I feel honored to be a Yisrael.

While Dad was in Germany, Allied soldiers liberated Jews enslaved in Nazi concentration camps. This echoed the biblical Passover story when God freed Jews from Egyptian bondage. When the commemorative holiday approached, Dad organized a small Passover seder for Jewish soldiers. "I prepared for about twenty men. However, hundreds of servicemen came. They all wanted to celebrate Jewish freedom in Nazi Germany. We hastily set up a loudspeaker and they stood outside listening as I chanted from the ancient Passover Haggadah."

When it was Dad's turn to come home from war-torn Europe, he sent Mom the departure date and the return vessel's name. But he didn't arrive when expected. Mom contacted the Army and asked if his ship landed. "Yes," they said. "The ship arrived as planned. The men have access to a phone. Your husband can call you whenever he wants."

This caused Mom to panic. She called hospitals and morgues. Two weeks after Dad's ship should have docked in New York, Mom still waited. The Army kept saying, "You're not the only wife who didn't hear from her husband yet."

Eventually, Mom learned from the radio that a terrible storm prevented Dad's ship from landing. The Army had reported my father's arrival incorrectly and they called her back with the news of their mistake. When Dad finally arrived at the Brooklyn apartment, Mom broke out in tears of joy. Diane didn't recognize Dad and didn't understand why my mother was crying. To protect her, Diane stomped her foot and yelled at the stranger, "Don't you make my mommy cry!"

When Dad came home from the War, he moved into Grandma's Bedford Avenue apartment, joining his wife and daughter. "Superior Printing Company" and I were born about a year later, the same year that Israel became a state. The goal

CHAPTER FOUR

of Superior, as Dad and his partner Marvin called their business, was to let them earn the money to send their children to summer camp and private college. However, having survived the Depression, the men were hesitant to resign their tenured school positions. They arranged to both keep their teaching jobs by working different school sessions, allowing one of them to always be at the print shop. Dad was to be the inside man, using the experience he gained from working at his uncles' printing business. Marvin used his affability as the outside man, bringing in the customers. The Jewish Welfare Board, to which Marvin's Army chaplain belonged, became Superior's first customer. Both the business and partnership were enormously successful for many years.

Birth and death are two sides of the same coin. At birth, we separate from the cosmos and God to exist as a soul on earth. In death, we meld back into God's bosom. In my family, a mistake in our DNA inherited at birth often encoded death.

It was Saturday, June 3, 1950. My mother, Cyrilla, threw a family party to celebrate my father's thirty-sixth birthday. He had been home from the Army for three years. I was two and my sister was seven.

Dad's mother, Grandma Marion, whose husband David died twenty years earlier, rang the bell at 5 p.m. Grandma Jennie opened the door and in a loud voice enunciating each syllable said, "Greetings and salutations!" This was her standard welcome. Since she came to the United States as a small child, she spoke without an accent. She was prideful of that. "Congratulations on your son's birthday."

Marion's broad smile thinned her eyelids as her cheeks rose and revealed her straight teeth. She had a heavy Russian accent. "Hi, Jennie. It's a glorious day. Can I help prepare anything?"

Jennie adjusted the pins securing her thin, snow-white hair in a bun and pointed to the dining room. "Cyrilla has everything laid out. Just get a plate and serve yourself."

When Marion saw the spread, she fanned the napkins, ordered the silverware, and rearranged the cold cuts, vegetable slices, and sour pickles. She made everything look festive in a few minutes. Mom sulked upon seeing her mother-in-law belittle her preparations, but she quickly masked this response with a smile.

Dad's brother, my handsome twenty-six-year-old uncle Cyrus, his wife Pearl, and their son Marty were also at the small party. Like my parents, Pearl and Cyrus met at Erasmus Hall High School. My cousin Marty and I were both terrible twos. "Greetings and salutations," Grandma Jennie boomed their welcome.

Everyone munched on the tongue, corned beef, and fresh Ebinger's rye bread while Diane performed for us in a tutu and pink toe shoes. Mom played the music for her ballet solo on our phonograph. Diane was a talented dancer, and the adults applauded as they balanced their plates and glasses of ginger ale, cream soda, or root beer. Marty and I ran around chasing each other until Grandma Marion caught us. She put us each on a knee to read us a story. Diane changed and sat next to us, leaning her head on Grandma Marion's shoulder. When Marty and I escaped, Grandma Marion played checkers and other games with Diane. Later, she entertained Marty and me with a red toy car as she made motor sounds that made us laugh.

Mom brought out the bakery birthday cake with all the candles. After we sang Happy Birthday and Dad blew out the candles, Grandma Marion stood up and rubbed her temples. "I'm sorry to leave so soon, but I have a headache. I think I'll go to bed early. Happy Birthday, Norman."

Marion slipped on her sun hat and hugged and kissed everyone good night. "See you tomorrow, Cyrus, when I pick Marty up."

Grandma Jennie walked over to Dad and used her theatrical voice to say, "On your birthday I wish for you everything

CHAPTER FOUR

you wish for yourself that is good." Dad smiled and shook his head, encouraging her to continue. He knew what was coming because he had heard it before. "Sometimes we mistakenly wish for things that are bad for us. I don't wish for you anything that you wish for yourself that is bad."

Dad and his brother Cyrus finished the evening amusing the kids and discussing current events. Dad devoured news about "Operation Wings and Eagles," which was just winding down. He scratched his head and pointed to the newspaper. "Israel finished airlifting 49,000 Yemenite Jews to their new state." He squeezed me and continued. "They say these Yemenite Jews lived in isolation for so long that they didn't know the Romans destroyed the second temple. Learning that was a terrible shock for them."

Cyrus scooped his son Marty up and put him on his shoulders. "Can you imagine having such a huge cultural shift in your lifetime?"

Dad wrinkled his nose. "I've been thinking about that. Despite their living without electricity or running water and not knowing about airplanes, their love of life, family, Torah, and God shows how similar we are."

Aunt Pearl, a redhead, joined the men and, pointing to Marty, said in her soft rasping voice, "We'd better take this little guy home." Then she turned to face my father. "Happy Birthday, Norman, and many more."

The next day, Marion didn't show up when expected to take her grandson Marty to the park. She also didn't answer the telephone when Cyrus called. Concerned, he went to her apartment. When she didn't come to the door, he let himself in with his spare key. "Mom, are you here? You forgot about taking Marty to the park."

When my uncle didn't find her in the kitchen, living room, or dining room, he entered the bedroom. There she was, still asleep. When he couldn't wake her, he panicked and called for an ambulance. The coroner ruled she had had a fatal heart attack during the night, but there was no autopsy.

Mom said Dad stayed in bed crying for a day when he heard the news. Marty remembers his father crying too. My sister Diane felt it as a terrible loss and still spoke to her son Jeffrey of the trauma thirty years later.

When I was old enough to understand, Dad gave me the impression that Grandma Marion died from a heart attack in old age. Although she was only fifty-nine, Dad implied this was a normal lifespan. I accepted this. Only after my niece's death did I recognize it for what it was: premature sudden death. If the family sudden death mutation came from my father, did he inherit it from his mother, Marion, or was it from his father, David, who died mysteriously at forty-two?

June 3 commemorates both my father's birth and my grandmother's death. Marion's soul had gone back to God's bosom.

We continued to live with Grandma Jennie until I was four. Herbie married and moved away with his wife and son, leaving my sister, Diane, and me as the only children in the house. Diane was my coach, mentor, protector, and loyal supporter. On Saturday and Sunday mornings, we listened to radio shows together, and when the family got a television we watched cartoons. Whenever our parents went out, we snuggled on the couch and stayed up late watching *My Little Margie*, *Father Knows Best*, and *Dragnet* and enjoying ice cream. Diane also concocted make-believe games that I never tired of playing.

When Grandma Jennie was sixty-three, my parents decided to move out of her apartment to get more space and privacy. But Grandma insisted on coming along. She told my mother, "I took you in during the War so it wouldn't be fair for you to leave me alone now."

CHAPTER FOUR

Grandma Jennie playing piano. Charcoal sketch on newsprint from life by me in 1960, when Jennie was seventy-three and I was thirteen. AUTHOR CREATED

My parents bowed to Grandma's wishes and still achieved privacy by renting a flat with a suitable floor plan. They chose an elevator building on Ocean Avenue seven blocks from the old walk-up apartment. There was a live-in maid's bedroom and bathroom that were far from the other bedrooms. These rooms were perfect for Grandma. My sister and I, then five and ten, each had our own bedroom near our parents' master at the other end of the apartment.

Because of its importance to Grandma, my parents kept a kosher home. They had separate dishes for meat and dairy meals and obeyed the special dietary laws followed by observant Ashkenazi Jews. Periodically, I would see silverware sticking out of a pot of soil. This was how Grandma purified a dairy utensil mistakenly used for meat.

During the years Grandma Jennie lived with us I don't remember her ever being sick in bed. She was an aloof, proud lady with her own friends and social life. Although she often exchanged angry words with my mother, Mom admitted, "Grandma is much easier to live with now that she is old." My father had a good relationship with her. He kept the peace. For decades he chatted with her as he put nightly drops in her eyes to treat her glaucoma.

Grandma contributed to the family chores by doing the food shopping. There were stores of all sorts on nearby Flatbush Avenue and Church Street. In her sensible lace-up black shoes with short thick heels, Grandma walked to a kosher bakery, butcher, and fish store and stopped at the kosher deli and general grocery store. Now and then I came along on these errands. Grandma wasn't tall, but she held her head high and kept her back straight. Sometimes she wheeled grocery bags home in a cart. Often, she carried one bag with just a few items because she shopped frequently due to the limited size of our freezer and refrigerator. Thanks to Grandma, we always had fresh produce, bakery goods, and meats.

Grandma Jennie ate with us every night, taking part in the dinner conversation but never monopolizing it. After supper she mumbled Hebrew prayers using a frayed Grace After Meals pamphlet that she peered at through her thick eyeglass lenses. Six evenings a week, my mother prepared dinner and cleaned up.

CHAPTER FOUR

For Shabbat, Grandma took over the kitchen to cook us a special meal. She made fricassee by boiling chicken wings and necks with the skin attached along with small greasy meatballs all in one pot. I was a picky eater in those days and dreaded this dish. Sometimes she made kreplach with kasha, which I also disliked. I viewed this as pasta ruined with oatmeal-like cereal. But I loved it when Grandma fried couscous with onions. This smelled and tasted fantastic.

Friday evenings Grandma Jennie lit Shabbat candles alone in her room before dark using plain bent metal candlesticks. She had a schmatte over her balding head of white hair as she muttered the traditional Shabbat prayer. I don't know why the family didn't do this with her. Our children, and now grandchildren, gather around when we light candles to welcome Shabbat.

Grandma didn't play children's games, but she taught me how to play *her* favorite game, Canasta. One afternoon she let me play it with her and her friends at our big dining room table. She didn't take me to the park or on trips but came to all of my school graduations wearing a proud smile. Occasionally, she took me to services at Congregation Shaare Torah a short walk from our house, where the congregation remembered my grandfather Simon as a revered scholar. She also sometimes sat in the green velvet lounge chair next to the piano in our formal living room to keep me company when I practiced.

FIVE

Self-Discovery

"For my sake was the world created." —Mishnah Sanhedrin 4:5

Judaism teaches that we are each unique and each have a special role; it is our job to discover what our mission is and to perform it. Good parents help their children in this task by celebrating their unique talents, which is how my parents led me into a life of science.

When I got home from school in second grade Mom was usually leaning on her headboard in her twin bed with her legs stretched out. She would be wearing her work clothes but not her shoes. I see her holding the rotary telephone handset sandwiched between her ear and shoulder as she sewed on buttons and mended hems. Instead of scissors, she used her teeth to cut the threads. I lay down next to her for a while. "Yes," she said, then there was silence for a long time. "Really," and more silence, then "Ah ha." This repeated itself with a series of callers. When I asked what they spoke about, Mom said it was private. Pearl, Alberta, and Edith just needed someone to listen. Mom never broke a confidence.

CHAPTER FIVE

Soon she would ask about me. She was interested in the minutia of my life. Even after I gave her the details of my unremarkable day, she felt cheated because I balked at repeating each story endlessly. "Tell me again what he said," Mom would ask eight or nine times. Only a video recording of my entire day would have satisfied her! I believe the reason for this compulsiveness was that her job was not fulfilling, so she tried to live through her children. Repeated incidents like this inspired me to work for a meaningful career and become a less smothering mother.

Sometimes I cut the cross-examination short by going off to my room to do my homework. One day I found a bag of new books waiting for me in my room. At seven, I was an avid Nancy Drew fan and was proud of the over fifty Nancy Drew books I had on my shelf with their dark blue covers and orange titles. I loved to read about Nancy solving mysteries. But I also enjoyed nonfiction books. My parents gave me a subscription to the Signature biography series. The life stories of Louisa May Alcott, Madam Curie, Clara Barton, Good Queen Bess, Amelia Earhart, Thomas Edison, Florence Nightingale, and Louis Pasteur inspired me. Many of these real "characters" solved problems and mysteries like Nancy Drew did. My parents also gave me nonfiction books about the establishment of the State of Israel.

That afternoon the new books in my room weren't Nancy Drew, biography, or history. They were about science: the weather, frogs, and electricity. I didn't understand why Mom had picked out this set of weird books. But I started reading them and lost track of time until the phone rang at 6 p.m. Dad called every night just before the rest of us sat down to dinner. His routine was to get up in the dark at 5:30 a.m. while everyone else in the family slept. He quickly slipped clothes onto his tall thin frame, combed his short brown hair parted on the side, and boarded the subway in time to teach his early morning classes. Then he changed clothes and hurried to his other job at his Superior Offset Lithography shop, where he remained until after 7 p.m.

SELF-DISCOVERY

Hearing Dad's call, Mom stopped dinner preparations and ran to answer the phone like a teenager in love. She picked up the hall extension on the small mahogany bookcase that held the Great Works and a miniature encyclopedia my father got when he became a bar mitzvah. When I was between five and eleven, I usually lay on the tapestry-covered couch in the living room watching my favorite TV show, *Howdy Doody,* when Dad called. From there I could see Mom lost in conversation on the hall chair.

The science books so engrossed me that evening that I forgot to watch my show. I was upset because I missed a serial within the program. My bottom lip protruded in a pout. "Daddy, I missed Howdy Doody tonight, and I don't know what happened to Princess Summerfall Winterspring."

Dad commiserated. "I am so sorry that happened to you."

If only I could turn back time, I thought. "Can you help me?"

Dad was an expert at making things better. "I'll call the TV station and ask them to replay the episode for you." This cheered me up. In my mind, Dad had the power to make that happen. He telephoned a few minutes later. "I reached the station, and they told me there was a technical problem. The show wasn't on tonight after all!" I can't remember if I believed this, but I know it comforted me that Dad listened and tried to make me feel better.

After the call, my mother, grandmother, sister, and I crowded around our tiny kitchen table for dinner. The apartment's architect designed the kitchen for cooking and easy access to a large connecting dining room. Earlier tenants separated these two rooms with a wall, making space for a small kitchen table sandwiched between the refrigerator and stove. We had my favorite dinner, rib lamb chops. I wasn't happy about the canned peas and carrots though. "Why did you get me those books?"

Her answer surprised me. "I read in the *Sunday Times* that girls are often afraid of math and science. I didn't want that to happen to you and thought reading about science would help."

CHAPTER FIVE

Why wasn't science just as suitable for girls as for boys? I wondered. Indeed, some of the scientists I read about in my Signature books were women. I tilted my head and narrowed my eyes. "Why would girls be more afraid of science than boys?"

Mom shrugged her shoulders. "I don't know. They shouldn't be."

"Thanks for the books." They were okay, nothing earth shaking. What was awe-inspiring, though, was Mom's interest in making sure I didn't shut the door on science and math. Although I was a girl and although she wanted me to get married, have children, and teach in elementary school, she also wanted me to give science and math a chance. This was big!

It would be a long time until I became a scientist and an even longer time until the family mutation reared its ugly head. Meanwhile, we unknowingly armed for tragedy by living each day and loving each other.

When Dad came home from the print shop, still radiating cold from being outside, he'd sit on the edge of my bed. I always waited up for him. He spent about fifteen minutes talking with me and patting my back in the dark while Mom broiled his dinner. This was our special time and my favorite part of the day.

"How are you?" he said.

I let go of my lovie doll Margie and turned on my side to face Dad. "School was okay, but Jane and Cindy didn't let me play with them again."

Dad stroked his chin. "What about Nancy?"

I swung my arms over my blankets and giggled. "Nancy always plays with me."

Rubbing his hands together and smiling, Dad said, "Nancy is a good friend. Should we ask Mommy if you can invite her over?" Dad always had great ideas.

I stretched out my arms. "Okay, I hope she can come."

Now it was Dad's turn to tell me about his day. "I learned something important from one of my students today. It's still on my mind."

This caught my attention. I moved Margie's arms up questioningly and asked in her voice. "How can a student teach a teacher?"

Dad answered Margie, patting her head. "I learn from my students all the time." Speaking to me, he said, "You know how I like to tell you, 'That's my girl!' when you succeed at something?"

"Yah." Dad said that to Diane and me a lot while clapping his hands to congratulate us. It made us feel good.

"Well, I say that to my students too when they do a good job. Today a student in my class conquered the use of a difficult machine after many failed attempts. I clapped my hands in joy and cried 'That's my boy!'"

But instead of taking this as a compliment, he told me, 'Mr. Weiss, I know you didn't mean anything bad by that, but please don't call me boy.'"

Dad then carefully explained the history of bigots insulting Black adults by calling them "boy" or "girl," and said, "I won't ever do that again."

Adjusting his metal-rim glasses and massaging the indentations near the bridge of his nose, Dad embarked on another topic, trying to give me perspective on my troubles. "You know, Susan, some children have real problems like not having enough food, not having parents, or not having a decent place to live. If you think about this, your worries won't seem so bad."

"But, Daddy, that's not true for any of my friends," I said, as our matching sets of hazel eyes locked. I mistakenly thought that all my classmates had a family like mine that mirrored the TV sitcom *Father Knows Best*. In this show, Robert Young played a wise, loving father who was a good breadwinner and happily married to a devoted mom, played by Jane Wyatt. Now I know fate didn't bless all my classmates in that way.

The next day my father left early to go to the print shop. As usual, on Saturdays, he returned home before lunch. There was a dusting of snow, so I talked him into dragging me around on a sled in the afternoon. He had loved sledding downhill on his Flexible Flyer as a boy in Pennsylvania. The metal runners on my Flyer weren't as slippery on a Brooklyn sidewalk! Still, Dad used his brawn to give me a ride.

When we came in for dinner, Mom gave Dad her place at the kitchen table and sat off to the side on a folding stool. She supported her plate on a low, free-standing metal cabinet across from the stove. On the wall just above Mom's stool, a half-door opened onto a manual dumbwaiter shaft. Our building superintendent called up to us every night when it was our turn to put the kitchen garbage on the rope-operated dumbwaiter. This was scary but exciting. We only used the dining room when there was company because it was much harder to serve and clean up if we ate there.

During dinner, Diane grumbled that a girl in her class teased her. Mom told her to ignore it. She grinned and said with a laugh, "If someone spits at you, hold out your hand and tell them, 'I think it's raining!'" Mom said that petty arguments occur most often during childhood and early marriage but that later things would be less contentious. "This too shall pass," she counseled.

That evening I confronted Mom with crossed arms and raised eyebrows. I was annoyed that there were four Susans in my class. "Mom, your name, Cyrilla, is so beautiful. Why did you call me Susan? So many girls have that name. Why didn't you use a little imagination?"

Mom ran her tongue across her front teeth and shook her head. "You know, Cyrilla is a lovely name now that I'm an old lady. But when I was a little girl, they called me Cyrilla, Gorilla, Sarsaparilla! Would you like that?" This made me realize Mom

SELF-DISCOVERY

still suffered from childhood taunts and that she named me Susan to protect me from teasing. Since then, I have loved my name and chose the simple names Michael and Judith for my children. I'm honored that my Michael married a Susan.

Mom was careful to avoid hurting anyone else's feelings and was sympathetic to others. Unlike her grandfather Victor, Mom loved people but was indifferent to causes that claimed to help mankind. She was extremely loyal to her friends and found great comfort in her family. Despite her difficult relationship with her mother, she let Grandma live with us for twenty-five years.

I was part of the Baby Boomer generation. Throughout the country, soldiers returned from World War II, after years away like my father, and had babies as soon as they got home. Because of the tremendous increase in the number of children, there weren't enough schools for us. To cope, schools created large classes or split the day into two sessions with half of us going to school early and the other half going late. Another way of relieving the overcrowded schools was to graduate students quickly. If you passed a test, they put you in a "Special Progress" (SP) program that completed the 7th, 8th, and 9th grades in just two years. I landed in such a class, but my two best elementary school friends did not.

When I entered Walt Whitman Junior High School in 1959, there wasn't a single familiar face in my class. Most of the other students knew each other and talked about parties they didn't invite me to. My sister solved this problem by convincing me to invite kids from my new class to my house for a party. It worked and I became one of the gang. More importantly, I made a close friend from my class, Rosie, who also didn't know the other kids.

Just before I turned thirteen, my parents started arguing in front of me for the first time. The United Federation of Teachers

CHAPTER FIVE

called a strike for November 7, 1960. Although it was illegal for schoolteachers to strike, Dad felt obliged to go along with it. He didn't need a pay raise, because his print shop supplemented his income, but many of his colleagues couldn't support their families with their meager teacher salaries. The superintendent threatened to fire every teacher who didn't show up for work. My mother feared Dad would lose his coveted positions as teacher and department chair if he didn't cross the picket line. She wanted Dad to focus on keeping his job to provide for his family, but she couldn't get him to change his mind. The same arguments and threats appeared twenty years later when the air traffic controllers called a strike in 1981, the year after my father died. It seemed improbable to everyone that President Reagan would fire all the controllers, but he did. That's when I realized my mother had had reason back then to be worried about Dad's job.

In the 1960s, the dilemma played out in households throughout the city, destroying friendships and unsettling families. Mom's brother was also a New York City teacher who supported the strikes. He described those who crossed the picket lines as traitors and ostracized them when the strike ended. Widespread reactions like this forced teachers to change schools or jobs. This was the first of a series of school strikes to disrupt my family.

No matter how we try to live our private lives, sometimes circumstances force us to take a stand.

Dad's other job, Superior Offset Printing, was in a sixteen-story building in the Turtle Bay neighborhood of Manhattan. One Saturday Dad took me to see it. We rode a large freight elevator to their loft with four-color presses, a wide assembly of paper-handling machines, an office, and a darkroom. I saw and smelled ink and paper, which were lying everywhere. Dad explained that his employees typeset, proofed, printed, sorted, folded, bound, and mailed each job in record time. The people hubbub and machine racket present on workdays were all missing that quiet Saturday, but I was still proud to see what my father had created.

We easily got seats in the mostly empty car on the train ride home. Dad told me about one of his policies as a teacher. "I always ask two female students, never one, to help me with clerical tasks. And I am careful to leave the door open." My father adjusted his jacket and rubbed his neck. "Both of these things are important to allay any question of sexual misconduct." Sixty years before the "Me Too" movement, Dad had figured this out. Both as a teacher and businessman, he followed the Jewish principle of *marit ayin* to remain above suspicion by avoiding all appearances of wrongdoing. In my school and professional interactions, I hold everyone to this standard.

Family was everything to my mother. For years, Mom's cousins met monthly in each other's homes. When it was our turn to host the group, they filled our living room with stories and laughter. Although we had a cabinet stuffed with unopened bottles of fine wines and other alcoholic beverages that clients gave Dad as Christmas presents over the years, my parents never drank or served these. I don't recall anyone drinking coffee or tea either. Instead, the cousins enjoyed black cherry soda, cream soda, root beer, and orange soda and ate nuts and chocolate.

Mom knew of families that fell apart over disagreements about parental care, finances, or other slights. She used to say, "Herbie is my only sibling. I am not going to let anything cause me to lose him." If Mom felt annoyed that she carried a larger burden of care for their mother than Herb, she never voiced it. We often visited with Herb, his wife, and their three children.

We saw less of my Dad's brother, Cyrus, who eventually divorced and moved to California. Mom used to remind Dad all the time—"Call your brother." She explained that another

CHAPTER FIVE

way families fell apart was by neglect, saying, "If you don't make an effort to keep in contact with Cyrus, you will eventually become strangers."

My family helped me understand who I was and discover what I was meant to do. To aid me on this journey, my parents exposed me to many different experiences, opportunities, and extended family. They showed me what fulfilled lives look like, which gave me the will to emulate their success. Even though they have been gone for many years, I can still touch the neverending well of their unconditional love, which continues to arm me with confidence and optimism.

It was the end of summer and I had come home from Buck's Rock sleepaway arts camp. I was fourteen and could hear Grandma Jennie's hearty laugh as I stood outside her door. A long narrow hallway separated Grandma's room just past the kitchen from the rest of our bedrooms. Her television was blaring because she was hard of hearing. She didn't notice my knocking until I banged with all my might. "Hi, Susan. Glad to see you're home," she said. "Join me to watch *You Bet Your Life*."

It worried and embarrassed me that everyone in the courtyard could hear Grandma's show through her wide-open window. I felt the welcome breeze in my hair as I sat down on the bench beside her and tried to make sense of the humor in the program. Groucho was telling a guest, "I see you got your looks from your father. Is he a plastic surgeon?" Although Grandma laughed at this, it made me uncomfortable. I thought it was mean. Now, viewing episodes as reruns I understand their appeal.

Grandma displayed the pottery I made for her at camp the year before in her room. She used my irregular turquoise tray to hold her false teeth overnight. I appreciated Grandma's interest in my clay pieces.

SELF-DISCOVERY

When the show ended, Grandma looked at me through her thick lenses in a gray translucent plastic frame. She needed thick lenses following her cataract surgery. "Did you enjoy camp?"

Blushing, I said, "I had two boyfriends this summer. We went to movies on the lawn and camp shows together."

Grandma sat up tall. "Now that you are dating and Diane is of marriageable age at nineteen, I hope your parents understand how important it is to investigate any liaisons."

My eyes widened. "What is there to investigate?"

"You have to find out about the boy and his family. My parents learned that my suitor's uncle had a disease that runs in families, and they didn't let me see him anymore." I already knew that Grandma expected us to marry Jews, but now I learned there were other requirements. She continued, "When choosing a husband, it is important to check his family's medical history. You should reject anyone with a family scandal or defect that might cause disease or mental impairment."

"Okay, Grandma I'll keep that in mind. Did Mom tell you she met cousins from your brother's branch of the family when she visited me on Parents' Day? Their son was at the camp too. He was quiet and I didn't know him very well. Mom was excited to see them because they don't come to the cousins' club."

Grandma Jennie lifted her chin. "Yes, she told me. She said they seemed surprised that you and Diane could go to that fancy camp."

"Oh. They didn't talk long. But now let me tell you more about camp." I clapped my hands. "There was so much to do there. I loved working with clay." Sculpting figures to support the weight of body parts was a challenge. It seemed like magic when glazes changed colors and became transparent and shiny upon firing. "This summer besides sculpture, I went to the painting and the silversmith shops and used a potter's wheel in the ceramics workshop to make a tea set for Diane. This took a long time because the same glaze on the matching pieces came out different depending on the heat level and duration of firing. I learned that I had to fire all the pieces at the same time if I wanted them to match."

I reached into a bag I was carrying and pulled out a present for Grandma. "Here is a wooden bowl I made for you on a lathe in woodworking."

Grandma beamed and squinted as she examined the bowl. "This is beautiful, Susan. Thank you so much. What type of wood is it?"

"It's walnut."

Grandma leaned forward and rubbed her hands together. "What else did you do besides art?"

"Some evenings we sang folksongs and there was folk dancing too." I still love folk music and folk dancing sixty years later. "And I got a small speaking part in a play, *The Life of Man*, by Leonid Andreyev. Of course, Diane was in lots of ballet performances."

I never forgot Grandma's warning about health considerations when choosing a husband. I now realize this attitude has caused families to hide and deny illness for fear marriage partners will shun them or their children. However, all families have strengths and weaknesses, and hiding a vulnerability does not make it go away. Had Grandma Jennie known of the secret surrounding the death of my father's brother Eugene, she might have objected to letting her daughter marry Dad, and my sister and I might not exist. Luckily for my mother, my sister, Grandma Jennie herself, and me, Grandma Marion kept her secret well. However, the cost of keeping that secret may well have been my niece's life.

SIX

The Packed Bags

> "Life is not measured by the number of breaths we take, but by the moments that take our breath away." —Unknown

When Diane was in high school, she woke up one day with a large ugly pimple on the tip of her nose. She couldn't face going to school until it receded and was sure Mom would understand. "I don't have to go to school today with this ugly pimple, do I?" she asked.

Mom's immediate reply was, "Of course, you have to go to school, Diane. School is not a beauty contest!"

Diane was unusually popular even though, or perhaps because, she had inherited Dad's severe acne. Mom said Diane's skin would have worried her a lot if the kids hadn't liked her so much. The key was that she learned not to let her bad skin affect how she felt about herself.

As a preteen and teenager, Diane's chief passion was boys. Boys liked Diane because she was vivacious, always saying something witty or interesting. They could just listen, and she would keep the conversation going forever.

CHAPTER SIX

Mom was also popular with boys in her youth. She loved reliving this experience through Diane and put her ear to Diane's door whenever she entertained a boy. Mom did this even though Diane would soon give her a detailed accounting. She just couldn't wait that long. But, as high school continued, Mom got increasingly concerned that Diane might not get into college because her interest in boys prevented her from doing her schoolwork. This led to screaming matches between them. As a result, my mother would often implore me, "You won't ever do this to me, will you?" This motivated me to become a diligent student. Despite Diane's lack of studying, colleges did admit her.

Pointe shoes in the hand are graceful. On my sister's feet, they were magic. When she danced with them, she seemed to float. I watched in awe. She created beautiful curves and shapes with her body.

Because of her athletic talents and dislike of academics, Diane enrolled in Bouvé-Boston, part of Tufts University, to become a gym teacher. She moved into her dormitory when she was still sixteen and I was in junior high. I missed her dreadfully. We couldn't talk much because interstate calls were expensive and charged by the minute (email and texts wouldn't come into existence for decades).

Instead of exchanging letters, we communicated through our parents. They served as operators telling us about the other sister. Since our parents filtered this information, we couldn't gauge for ourselves how the other sister felt. And we couldn't provide each other with congratulations, comfort, or advice. While we were close during childhood, throughout my adolescence and entrance into adulthood, our relationship became distant. I began to see Diane more as an important character in my parents' lives and less as my sister. When she came home on school breaks, my anticipation was so intense that I broke out in hives. I couldn't wait to hear all about her adventures from her own lips.

During her freshman year, we were surprised to learn that many of the courses needed for her physical education major were also required for premed students. Being in competition with premed students was untenable for Diane. She solved the problem by transferring to the Eliot Pearson early childhood division of Tufts. Our parents encouraged this because they thought working with little children was easier than dealing with junior high or high school students.

Diane's focus on boys and dating continued throughout college. She also spent time learning and playing bridge, which she continued to enjoy throughout her life. The social scene was perfect because of the many colleges in the Boston area. Diane dated lots of boys at once. She loved wearing stylish clothes with sexy high heels. She had boundless energy and rarely complained of pain or exhaustion until the end of her life.

However, Diane felt insecure with academics and saw herself as a poor reader with no interest in books. She rejected one Harvard suitor whom she had dated for years because she feared he would regret having a wife who didn't measure up academically. She also worried that she would feel inadequate. Surprisingly for someone who focused on boys over academics, Diane graduated from college on time, but without a ring or steady boyfriend.

In the spring of Diane's senior year, 1963, she came home to join us at the family Passover seder. Dad was a few months shy of fifty then. This birthday was a landmark for him because when he was a student teacher in his twenties a fifty-year-old colleague in his school suddenly died. The entire staff was in shock at the young death. For Dad, however, with the pain of his father's passing at forty-two still raw, living to fifty seemed like a satisfactory span

CHAPTER SIX

of years. His statement, "At least he lived long enough to experience a full life," horrified the other teachers, and they rebuked him as an insensitive youth.

It was unusual for anyone to criticize Dad as heartless, so this incident stuck with him. He still thought about it, especially now that he was reaching fifty.

Every year, I looked forward to the seder when we celebrated our freedom from slavery in Egypt as described in the Torah. We put the leaves in the long oval table earlier to make room for my parents' friends, the Grosses. Then I retrieved our white china dishes with gold trim from the bottom of the breakfront. They were behind sticky doors, so I had to use all my strength to get at them. Kneeling, I handed the dishes up to Mom, who put them on the dining room table covered with an off-white tablecloth. At each place setting we added crystal glasses, sterling silver utensils, and an illustrated Haggadah booklet that prescribed the order of the ceremonies and listed the prayers and songs.

As the seder leader, my father sat at the head of the table with plates displaying Passover symbols and three ceremonial matzahs (unleavened bread) nearby. Mom was at the foot of the table, near the kitchen.

Dad led the service according to the order set out in the Haggadah. "Cyrilla, please light the candles." Mom, wearing a green shirtwaist dress with a headscarf shrouding her eyes, recited the blessing while waving her hands over the lights. The fragile flames flickering in Elke's silver candlesticks reminded us of how easily the light of freedom can be extinguished.

Wearing a suit and tie for the occasion, Dad poured sweet Manischewitz into his cup and passed the bottle. "Please fill your glasses," he said. Then he leaned forward and lifted his crystal goblet, chanting the Kiddush prayer over the wine before we drank it.

Diane removed her high heels as she relaxed in her chair and enjoyed her wine. She looked like an adult in her red sheath dress with puffed sleeves and eight large white buttons. I was a junior in high school and contented myself with a few sips of wine while trying to recline in my chair. Reclining on pillows was customary to remind us that we were free and not Egyptian slaves.

"I'm going to wash my hands now," Dad said. His white satin kippah engraved from Cousin Marty's bar mitzvah fell off as he hurried to the kitchen sink. Dad caught it and put it back on his head of deep brown hair, where it made a tent on his head. He returned to the dining room carrying a prepared plate of celery sticks and a bowl of saltwater, which he handed to his friend. "Please take a piece of celery, dip it in the saltwater, and pass it on."

Dad made a show of breaking the middle of the three ceremonial matzahs set out on the table. He wrapped the larger piece in a napkin. This was the Afikomen, which he would soon hide. Diane and I still loved to find the Afikomen and demand a ransom before returning it. This game of hiding the Afikomen is an Ashkenazi tradition.

Since I was the youngest at the table, it was my role to ask about the unusual ceremonies at the seder. I chanted the four questions prescribed in the Haggadah. I learned to sing these in Hebrew as a child. Dad began the answers by explaining, "During Passover, we don't eat any foods with leavening because we didn't have time for the bread to rise when we left Egypt." He then called on everyone in turn to read the story of our redemption from slavery. We read in English, but Dad added fun by speed-reading Hebrew when it was his turn. Dad made the descriptions of the four types of sons—wise, wicked, simple, and too young to know how to ask—more inclusive by converting sons to children. He made sure neither Diane nor I got the part about the evil child so we wouldn't feel bad.

We celebrated and whooped, singing Diyanu. The simple chorus of "Di, diyanu, diyanu, diyanu" means "it would have been sufficient," and it followed each verse about a wonderous thing

CHAPTER SIX

God did to help the Jewish people. Mom and I complemented the song's rhythm by tapping our spoons on the china and crystal.

Diane didn't take part. She frowned as we whooped. "I don't like this song. Di, di sounds like my name and like the word 'die.'"

Diane often pointed out hints of death and dying that didn't make sense to me.

As we recited the names of each of the ten plagues God sent to the Egyptians, we spilled drops of wine from our glasses onto our plates with our pinkies. After the last plague, Diane and I licked the wine off our fingers.

Lifting his glass, Dad said, "Let's drink our second cup of wine."

When he left to wash his hands again, as prescribed in the Haggadah, Diane twisted in her chair. "Where is the Afikomen? Does he have it with him?"

I got up and followed Dad as he went to the kitchen sink. I was looking for the Afikomen. "I don't see it."

Back in his seat, Dad said, "Let's each eat a sliver of matzah with a slice of horseradish to recall the harsh lives our ancestors had as slaves."

It was very bitter.

"Now we eat matzah with charoset." This was a paste of apples, nuts, honey, and wine I helped make earlier to remember the mortar we used to build the pyramids. It tasted much better than the horseradish.

These ceremonies complete, Mom stood and announced, "The festive meal will now be served."

Seders used to drag on forever, but tonight the service ended much too quickly. I yearned for more. "Daddy, what happened to the rest of the service? Why did you cut so much out?"

Dad pointed to his marked-up Haggadah, which showed the prayers and songs he skipped. He smiled and chuckled as he leaned back on his chair. "I've skipped exactly the same things for the past fifteen years. The only thing that changed is you." This was a vivid example of time speeding up as we get older.

Diane and I helped bring in the hard-boiled eggs from the kitchen. Special for the seder, we ate them cut up with saltwater. Mom's matzah ball soup came next, followed by first-cut brisket, gravy, potato pancakes with fried onion, and salad.

After we ate the chocolate and sugared fruit slices for dessert, Diane put on her shoes and leaped up. "Come on. Let's find the Afikomen." I followed. The Afikomen wasn't in the kitchen or hallway . . . not in the dining room bureau or large breakfront . . . not on the green carpet. And then I saw it—a little bump under the tablecloth near Dad's plate. Dad and Mom gave us costume jewelry for our find and we ended the meal by eating bits of the Afikomen.

A large glass of wine sat in the middle of the table for the Prophet Elijah. I delighted in the charade of inviting him to our seder, and I ran to open the apartment door to let him enter. Dad shook the table so it looked like Elijah was taking a sip, but the level of wine in the cup was unchanged.

"Why doesn't Elijah drink more of his wine?" I asked.

Every year Dad explained, "Elijah visits so many families that he can only taste a tiny amount in each home or he would get drunk."

This is when Dad tapped his wineglass to get our attention. He stood tall with his chin up and chest out. "Before we continue with prayers and songs, I want to share something with you." Dad looked at each of us. Referring to his long ago faux pas when he said his colleague who died at fifty had lived a full life, Dad said, "I'll be fifty in two months, so I can finally say it. I'm satisfied with my life. My bags are packed. I'm grateful for my wonderful marriage, children, and successful career. While I would enjoy having more years to see my children grow, I've completed my job with them. I'm ready to go anytime."

Two of Dad's grandchildren at the end of his life. Cousins Karen and Michael play together. They helped complete Dad's joy. FAMILY PHOTO

This statement stunned me. Why was he talking about dying? He was healthy and strong. I wasn't ready to contemplate losing him. His nonchalant talk of dying upset Diane. He must have told Mom what he was going to say ahead of time, because she remained calm.

The Grosses took it as an invitation to talk philosophy and joke. "Exactly where are you going with those bags?" they asked. To help us understand the impetus for his proclamation, Dad told the story of the day he said living to fifty was a full life when he was a student teacher after a fifty-year-old colleague suddenly died.

Fifteen years later at the Passover seder, when Dad was sixty-five, he reminded us again that his bags were packed. He cried as he told us how grateful he was that he lived to see his four grandchildren, Jeffrey (twelve years old), Karen (eight), Michael (five), and Judith (one). I didn't worry about these tears because it wasn't unusual for Dad to cry. He cried at the movies. He cried when he read children's stories. He was very sentimental.

Upon graduation, Diane moved back home and started teaching first grade. Now we weren't fearful about her career but that she missed her chance to find a husband! She tried going out with old boyfriends but wasn't interested in any of them.

Then she went on a blind date with Ronald Rothman, a tall, handsome Army lieutenant. Ronnie was a graduate of Dartmouth College, had blond hair, blue eyes, and an engaging smile. He loved sports and, like Diane, was athletic. They were an unbeatable mixed-doubles tennis duo. They married on December 15th, six months after Diane's graduation, when she was twenty-one.

As a new bride, Diane lived on Governor's Island, where the Army stationed Ronnie. She commuted by ferry to her teaching

CHAPTER SIX

job in Brooklyn. Diane spoke with Mom on the phone several times a day to share details of her new life. Upon returning from her honeymoon, she called to say, "When I started class today, I announced, 'My name is Mrs. Rothman.' One of my students raised her hand. 'Are you sure your name is Mrs. Rothman? Because you look exactly like our old teacher, Miss Weiss.'"

Ronnie's father, Charles, encouraged him to pursue a career in law. I'm not sure what Charles did for a living, but he convinced Ronnie that he had connections and would get him a good job if he became a lawyer. So, after Ronnie left the Army, he enrolled in law school. Sadly, at about the time Ronnie passed the Bar, his father, a heavy smoker in his late fifties, came down with lung cancer and died within a few months. Ronnie was bereft. In addition, he suddenly had to face the job market on his own. His excellent credentials led him to a secure an exciting civil service attorney position at the Internal Revenue Service.

Life was stressful for the young couple as they struggled to adjust to adulthood and marriage. My dad often served as their peacemaker. "As long as you are married, I must be impartial in mediating any dispute," he told Diane. "But if you ever separate, I will be 100 percent on your side."

SEVEN

Coming of Age

"Do your best and leave the rest!" —My father, Norman Harold Weiss

I rushed from English to honors geometry class lugging a heavy briefcase. The Erasmus Hall buildings were so large that it was often difficult to get from one class to another in time. But my English and geometry rooms were near each other so I arrived before the passing period ended. I slid into my assigned seat and began work on the daily extra credit challenge problem on the blackboard. As students arrived, they engaged with the problem silently. Mr. Deutsch stood in the front of the room. The small round indentation on his right cheek changed shape as he talked and smiled.

This was my favorite part of the school day. The problems required a deeper understanding of geometry principles than presented in our textbooks, and often they could only be solved by using novel approaches. When the bell rang to announce the start of class, those of us with a complete solution handed it in.

CHAPTER SEVEN

Mr. Deutsch called on Jerry Sussman to go to the board and explain the answer. Jerry was a wiz at math. This time I also solved the problem. I raised my hand. "I answered it another way."

"Great. Come and show us your method." Mr. Deutsch erased Jerry's work and handed me the chalk. I outlined my solution. "Nice approach, Susan." Mr. Deutsch's compliment made me feel valued.

For the rest of the period, Mr. Deutsch put proofs on the board, and I furiously copied them. As a lefty, I twisted my hand around so I could see what I was writing, which caused pain in my arm, wrist, and fingers.

Mr. Deutsch's class began my intense interest in mathematics. High school biology and chemistry focused on learning facts and definitions, which didn't capture my attention. Only when I understood science in terms of solving mysteries, did I come to love physics and biology.

After the last class of the day, I walked over to the Erasmus XYZ club, where I was a volunteer math tutor. Some students requested help on their own; others came because teachers referred them. I often worked with the same student for a week or more. Tutoring helped me see distinct aspects of mathematical principles. I simplified things to basic ideas and enjoyed seeing students gain confidence as they began to understand. I wanted to be a high school math teacher.

During my senior year, Mr. Moskowitz, a mathematician with a profound understanding of calculus, was my Advanced Placement (AP) calculus teacher. He lined us up against the blackboard to draft answers to his questions. He presented the material logically from different points of view so that we could understand the principles.

Partway through the year, Mr. Moskowitz had to take a leave of absence for several months. The school employed a tall, young gym teacher to cover for him. He often made mistakes and couldn't answer questions. Until then, I had a low profile in school. I hid my grades because I thought boys wouldn't like me if my grades were better than theirs. In this class, I sat in the front row next to Alan Baum. We both liked math, and when the teacher made a mistake, we raised our hands to correct him. We also helped answer students' questions. I tried to be respectful, but it frustrated me when the teacher got basic things wrong. One of my girlfriends asked me why I was correcting the teacher; she thought I was doing it because I sat next to Alan and somehow learned of the teacher's mistakes from him. It didn't seem possible to her that a girl could catch these errors.

Every night, when Dad came home from work after 8 p.m. he found me completing an unreasonable amount of homework. I toiled and stayed up long after he, Mom, and Grandma went to bed. This stressed and exhausted me. I can still hear Dad advise, "All you can do is your best. Let the rest go. Mom and I will love you just the same no matter what grades you get." When I went to bed each night, I felt immense relief and joy at not having to study for all the precious hours until morning.

We repeated this scene nightly during my three years in high school. In the end, Dad's mantra became, "Do your best and leave the rest!"

To relax on weekends, Dad and I watched *Gunsmoke* and *Have Gun Will Travel* on a portable television in my sister's room, which she had asked my parents to furnish like a den. We ate coffee ice cream and reclined on her gray-and-white plaid studio couch. Diane was away in college then. Coffee flavor substituted for chocolate, which we avoided because they said it caused acne.

I didn't realize it then, but Dad had his own tensions. Mom told me Dad had rare but intense temper tantrums. Dad's younger brother, Cyrus, was also prone to tantrums and, according to his first wife, these occurred often. Possibly both brothers expressed pent-up fury at their father's early death.

Cyrus's second wife once asked Mom how his first wife could have let Cyrus go. In her eyes, Cyrus was perfect. Mom sheepishly mentioned the problem of temper tantrums. "Oh that," she said. "I took care of that right away. When Cyrus pulled that stunt on me, I told him if he ever did it again, I would leave him the next day. That was that. It never happened again."

Dad's personal struggles made him sensitive and compassionate. After he prevailed over his demons, he helped others by sharing his hard-fought recipes for happiness. I never could have succeeded in my career and definitely not in life without Dad's guiding hand. He taught me to appreciate the absence of disaster and to be satisfied with my best effort.

My mother enjoyed seeing me in new clothes because most of my wardrobe consisted of hand-me-downs from my sister. She insisted on buying me fashionable clothes. As a child I didn't want to interrupt playing, drawing, or reading to go shopping. Instead, Mom walked over to Macy's herself and brought home outfits in several sizes for me to try on. She returned the rejected items the same day. Despite my complaints, I enjoyed the clothes and always got compliments on them. I recall a teacher praising an outfit in amazement because the flower trim on my red knee socks and cardigan sweater matched.

As a teenager, Mom sent me to the beauty parlor regularly to get my hair cut and set. I inherited her genes for premature gray hair, and my first white strands showed up when I was eighteen. She suggested I color these, following her example. I'm not sure why I refused. Only after her death, when I was forty with salt-and-pepper hair, did I dye the gray.

Mom wasn't always interested in clothes and styles. Dad's young aunt, Sylvia, who grew up in his household when her mother died, wore an elaborate coiffure, fashionable makeup, and stunning attire. She tried everything to get Mom and Dad to be more stylish. "You look like squares," she mocked, pointing to Mom wearing a simple cotton dress with only red lipstick enhancing her beautiful face.

Mom's eyes teared as she told me of this long-ago ridicule. "She is just a clothes horse."

Uncle Cyrus's dapper hats and catchy bow ties showed Sylvia's influence, but she couldn't interest my father in clothes. He refused to buy pants with narrow legs even when wide legs were long out of fashion. "Susie, whatever the style, I need to wear trousers with loose legs so I can change between jobs without taking my shoes off."

We saw little of Sylvia when I was growing up. Years later, Mom admitted she was wrong about her. "Dad and I would have done better in life if we paid more attention to wearing stylish clothes." I don't know what Mom meant by this. They may have felt better about themselves, but considering their achievements in work and social life, I doubt they could have been more successful. More likely, Mom's conversion was to encourage me to pay attention to fashion and help me do better in life. This concern must have made Mom realize that Sylvia likewise wanted to be helpful, not hurtful.

After I left home, Mom still wanted me to dress like a chic New Yorker. She gave me bellbottom pants and ponchos as soon as they came out in New York City. But wearing these clothes made me self-conscious; they didn't reflect my true self. This became a source of tension between my mother and me. She had become Sylvia and I the "square."

CHAPTER SEVEN

My high school AP zoology teacher, Dr. Lawrence, influenced me even more than my math teachers, Mr. Deutsch and Mr. Moskowitz. Dr. Lawrence changed my life. I took his class in my senior year. More important than teaching us zoology, he trained us to be confident in the face of challenges by teaching us perseverance and courage. He told of his lifelong battle with malaria. He encouraged all his students to enter the Westinghouse Science Talent search (later called Intel and then Regeneron Science Talent Search). If we entered the contest, he excused us from taking any examinations in his zoology class. Because of Dr. Lawrence, Erasmus Hall High School had spectacular success in this premier national science competition for over a decade.

The summer before I entered Dr. Lawrence's class, I took a college course in microbiology at Cornell University. This was part of a special program for high school students. We enrolled in summer school classes along with college students and adults. My course began each day with a one-hour lecture in a small, air-conditioned hall with comfortable movie theater-style chairs. It fascinated me to learn about the genetic material and that DNA had a double helix structure. After the lecture, we spent the rest of the day in the laboratory. My lab partner was a family man of about forty, but we worked well together.

Cornell reserved a dormitory on campus for the high school students. I lived in a nice suite on the third floor of a small walk-up building. My two roommates and I got along well. It was hot that summer so we had to open our windows wide at night to be comfortable. Since there weren't any window screens for the first two weeks of our stay, we suffered from mosquitoes.

We looked forward to a study break each night when the hamburger truck arrived. I craved their charcoal burgers with onions, ketchup, and pickles. On weekends, the program arranged trips to nearby parks. Cornell is surrounded by stunning countryside. I skipped from stone to stone in forest streams and delighted in beautiful waterfalls.

One detail I learned during this course was that taxonomists realized that two bacteria previously classified as distinct species were actually closely related and should be reclassified as one species. I decided to examine this idea in more detail for my Westinghouse Science project.

I asked my father to help me get the original journal article in which bacteriologists proposed the reclassification. He took me to a university library that was open to the public in Manhattan. The library was one large room with soaring ceilings and massive wooden tables. There was a mesmerizing, hushed ambiance. I found the journal issue I needed on the crowded shelves: it was bound in a huge, heavy volume. The article was too long for me to read in the library and they wouldn't allow me to check it out. However, they had a "white on black" copying machine available that cost just twenty-five cents a page. This was my first encounter with a copier. It seemed like magic that I didn't have to sit there for hours taking handwritten notes. Seeing a copier work was also enthralling for Dad as a printer.

The article convinced me to go ahead with the project and compare the two bacteria using all the laboratory tests I learned about in the summer course. I wrote to the Cornell professor and asked him to send me the bacteria. Once they arrived all I needed was the materials for each of the tests. I used a home closet as an incubator, and I sterilized supplies with a kitchen pressure cooker. Obtaining agar plates to grow bacteria on proved to be an obstacle. Every time I made these plates, mold contaminated them. My father suggested I buy plates already made. There was a laboratory supply house just a few blocks away from my high school, so I walked over there after school one day and inquired.

CHAPTER SEVEN

The business owner asked me why I needed the plates. I told him of my ambition to enter the science talent search. "Don't waste your time on that," he scoffed. "You don't have any chance of winning such a competition. It's not for the likes of you. You think you know what you are doing, but let me tell you, you are totally unprepared."

I left in tears, dragging myself home along Flatbush Avenue in humiliation. It was dusk, so the lighted signs above the stores were just turning on. They flashed in a multitude of colors. I passed the Five and Dime and Macy's stores along the way. "Come in and shop here," they seemed to call in competition. The familiar smells of pizza from the pizzeria, popcorn from Lowe's movie theater, chocolate from Ebinger's Bakery, and burgers from Charcoal Chef didn't distract me from my shame. How could I have been so naive as to think I could compete in a science talent search?

When I told Dad what happened, he clenched his fists. "What a jerk! Just ignore that idiot. He should be ashamed of himself for discouraging a conscientious student like you."

Despite Dad's long hours as a teacher and print shop owner, he made time to help me. He found another laboratory supply house in the yellow pages and called them to order the supplies I needed to complete my experiments.

In addition to submitting a research report on our experiments, the "Science Talent Search" competition required us to take a qualifying examination. Dr. Lawrence counseled us on how to deal with this and other competitive examinations. As he leaned against the laboratory demonstration bench in the front of the classroom, his thin frame and hunchback were striking next to the human skeleton on display. "It is important to study long before your examination so you can get to bed early the night before the test. No last-minute cramming." With his gaunt face and hollow cheeks, he flashed us a warm, toothful smile. "Remember to use all the time allowed to take any test. Never leave early." He had us spellbound. "And if you arrive before time, review material on the steps until the moment they admit you to the examination room."

I followed his guidance and went to bed early, but during the test, hives started appearing on one side of my body. Soon one eye swelled shut. While annoying, this wasn't that unusual for me during a stressful examination. However, I also felt sick. My throat was sore, my head ached, and I couldn't concentrate. By the time I got home, I had a high fever and lay in bed for several days. I thought that was the end of my chance at the science competition. But, surprisingly, I did well enough to qualify.

In late January, the Science Talent Search sent out letters announcing the 1964 Westinghouse semifinalists. There were 317 semi-finalists nationwide that year—the top 10 percent of all entries. Five of my classmates got letters telling them they were semifinalists. No such letter came to me. This disappointed me, but I became reconciled to my fate. Then the next day my letter came too. There was joy in Mudville!

Two weeks later, when we expected the finalist letters to arrive, my friend Andy went home for lunch to check his mailbox. He was indeed a finalist. Later that day I found a letter telling me I, too, was one of forty finalists nationwide! I knew Andy from grade school when we played hide-and-seek and sardines together outside. We continued the friendship in high school and throughout college. Not bothering with a coat or hat, I flew down Ocean Avenue to share the news with him.

In the following weeks, college admission and scholarship offers flooded my mailbox. Newspapers and radio programs interviewed me. Winning this honor was one of the most exhilarating events in my life. I became a different, more confident person. My parents were as excited as I was. They cut out every newspaper article, and Dad made offset copies of them.

The Science Service invited the forty finalists on an all-expense-paid trip to Washington D.C., to present posters of our

CHAPTER SEVEN

1964 Science Talent Search finalists on the Capitol Building steps. White arrow (middle) points to me, black arrow (left) to my classmate, Andy. AUTHOR-MODIFIED PHOTO OF SOCIETY FOR SCIENCE PHOTOGRAPH, REPRINTED WITH PERMISSION

work and compete for cash prizes. During our time in Washington, we had the opportunity to appear on a TV show. Additionally, the Science Service organized meetings between groups of us who were working on similar projects and renowned scientists in corresponding fields. My group of six had the privilege of speaking with Marshall Nirenberg, which was an extraordinary experience. While downing four cans of cola, he enthusiastically shared insights about his research and illustrated them on his office chalkboard. Four years later, Dr. Nirenberg earned a Nobel Prize for his groundbreaking discovery of the genetic code.

All two thousand students in my high school class were college bound. Because there were so many of us and the office didn't have word processors or copiers, we were each only allowed to apply to three private colleges plus a public option. The guidance counselors recommended we apply to a reach school, a safe school, and a fallback school. Since many of these colleges had a quota for the number of slots allotted to individual high schools, competition in my large class was fierce.

Because I was five years younger than my sister, I didn't start applying to colleges until she had already obtained her bachelor's degree. She had the radical idea that I should apply to the Massachusetts Institute of Technology. Having dated MIT boys, she knew there were girls there too and thought the male-to-female ratio of around twenty to one would be an asset for her shy sister. I was excited about MIT because of its stature in science—as well as the male/female ratio. I applied to MIT as my reach school, Jackson College as my safe school, and Cornell Agriculture College as my fallback school. I also applied to Brooklyn College.

After I handed in my college applications, I relaxed a little for the first time in years. While I didn't know if MIT would accept me, with the Westinghouse honor I felt confident I would get into Jackson College and Cornell. Going to either of these colleges would have been fine with me, so I forgot about the pending applications and had my first steady boyfriend, George. This romance taught me the importance of having a balanced life instead of working all the time. George and I socialized with four other couples that year. We had parties, went to museums, and went skating together.

I was thrilled when MIT admitted me, and my parents were delighted and proud. They were happy to have me continue to study math and science and have a chance to meet a genius to marry. They still assumed I would become an elementary school teacher. That wasn't to be. MIT didn't even offer education courses. None of us ever thought to check on that! But I think my parents would have let me go to MIT even if they had known.

From my high school class, Alan Baum, Jerry Sussman, and I all went to MIT. Andy went to Harvard. Alan became a University

CHAPTER SEVEN

of Michigan math professor and then had a lengthy career as a mathematician at General Motors. Jerry is an MIT electrical engineering professor. Andy became a physician like his father. I lost touch with my friends who went to other schools, including my boyfriend, George, who went to Columbia College.

In my sophomore year, my father sent a newspaper clipping about the owner of the neighborhood laboratory supply business who tried to discourage me from entering the Science Talent Search. I was shocked to read that a judge convicted him of supplying false laboratory results. This clinched it for me. Dad was right; the man who had so discouraged me was just an all-around jerk.

My parents expected me to go to college, become an elementary school teacher, and get my M.R.S. degree. They believed teaching was a secure job with good pay, good working hours, and lengthy vacations that were suitable for a mother. Even though my future husband was to provide the principal family income, my parents emphasized the importance of my being able to support myself and my family should the need arise. To guarantee this, they recommended I maintain an active career, ready to mobilize for full duty at any time. They pointed out that Grandma Marion's established dressmaker career was essential for the family when her husband died unexpectedly. I could see that Mom's work as a school clerk not only provided the family with extra security and income but was also an important social outlet for her.

I have often wondered why my parents gave me dance, piano, and art lessons, since they wanted me to be an elementary school teacher. Why did they enroll me in college board preparation classes and college summer courses? Why was it important for me to go to a good college?

They may have felt I would have a richer life with hobbies, friends, and non-career activities if I understood the arts and had a good education. Also, they may have thought my attending an Ivy League college would help me find a more successful husband. In the end, I believe their true goal, unknown even to themselves, was to allow me unlimited choices to reach my potential and be whatever I wanted to be.

Some scientists can point to a specific eureka event that signaled their interest in becoming a scientist. For others, including me, the realization and self-discovery was more gradual. When I was in elementary school, I wanted to be an elementary school teacher. When I was in high school, I wanted to be a high school math teacher. And when I was in college, I wanted to be a college professor. Eventually I realized that what I loved was not only teaching but also the thrill of learning something new and creating new knowledge as a researcher.

This is how I slipped into a career of university teaching and research. My parents supported me in this transition, relinquishing their dream of my becoming a New York City elementary school teacher.

EIGHT

The Dormitory

"Friends are as companions on a journey, who ought to aid each other to persevere in the road to a happier life." —Pythagoras

My mother persuaded me to leave Margie at home when I entered MIT. "They will think you are a baby and won't want to be your friend if you take that old doll." I was sixteen, well past the age I should need a security object, so I reluctantly agreed to put Margie away for safekeeping on the top shelf of the armoire in my room.

Margie is in my earliest childhood memories. As we played, I told her about my troubles. When I was angry, I yelled at her and threw her across the room. When I felt better, I hugged her soft stuffed body with movable cylindrical arms and legs covered in a thin blue plaid fabric and kissed her hard thin plastic face. She always forgave me and loved me no matter what I did.

By the time I was in grade school, my harsh treatment of Margie had taken its toll. Areas of her face were pushed in. Her stuffing was falling out and she was dirty. Mom wanted me to

CHAPTER EIGHT

abandon her. I had plenty of other dolls and stuffed animals, but none of them could replace Margie.

Although I refused to give her up, I agreed to let her go to the doll hospital. This was a mistake. They weren't able to repair her face. While I had expected them to just patch torn areas on her body, they covered her in a different material. The new fabric was thicker than the original and was a bolder blue with a larger plaid. At first it hardly seemed like Margie. It took time for me to love her again.

Margie came to camp with me every summer. She had even come with me to summer courses at Cornell University and Harvard College when I was a high school student. She never caused any trouble there. Now, as I left for college, she was alone on the top shelf of a dark armoire.

There were fifty women and nine hundred men in my MIT class of 1968. It thrilled me to be there. From my first day onward I thought about Dad's advice: "Do your best and leave the rest." MIT was much less stressful for me than high school, although many of my college classmates from less competitive high schools had the opposite experience. I was still conscientious about my work but found time for friends and extracurricular activities.

I met Razel Wittels the day I moved into McCormick Hall, the new women's dormitory. We spoke every night about boys, our experiences, worries, and dreams. During the day we were in different classes with separate circles of friends, so Razel was someone to confide in and listen to at night. She became family.

MIT gave me my first choice of room type, a single. The school also honored Razel's request for a double. They assigned her to a large corner room down the hall. The first week of school, Razel's roommate and a girl in another double on the floor exchanged rooms. Rumor had it, Razel's white Anglo-Saxon

Protestant roommate's mother couldn't bear having her daughter share a room with a Jew. Another family likewise deemed a Catholic girl unworthy as a roommate for their daughter. The solution was for the Jew and the Catholic to become roommates.

Razel and I liked folk dancing. We discovered and joined a community International Folk Dance group that met in a church in Cambridge. The group leaders wanted to move the dancing to a large ballroom, the Sala de Puerto Rico, in the new MIT student center. They encouraged Razel and me to start an MIT folk dance club and to invite the current dance instructors to continue to lead the dancing at MIT. We agreed, and it became an official MIT club with Razel and me as officers. MIT students stared at the dancing as they walked by and, little by little, they joined in. This dance club continued to meet at MIT for decades.

I took non-credit drawing and ceramics classes in the student center, and in my freshman year I performed in a one-act play. This is how I met my lifelong friend Ellen Greenberg. The director of the play was an experienced ballet dancer. My poor posture dismayed her, and she worked with me to reduce my swayback. I thought at age sixteen it was too late for me to correct my posture. I am very thankful that someone helped me with this while I was still young.

My dormitory family with Razel enlarged to include Ellen. The three of us supported each other with advice, criticism, and suggestions every night without fail. They sat on my bed, and I stood in front of my dresser putting my hair in curlers while we talked. I could see them in the mirror as I clipped the uncomfortable rollers in place.

After graduation, I remained in Boston for a year but then left the area when I got married. Razel and Ellen each settled in the Boston area. They continued to see each other regularly until Ellen, who studied linguistics at MIT and business at Columbia University, moved to Bulgaria on a Fulbright fellowship. She was a business professor at the City University in Sofia, Bulgaria, and founded and directed a medical center in Sofia. Ellen had many close friends but never married. Razel became a high school math and science teacher, after-school activities musical theater director, wife, mother, and grandmother. Although my ties with these women became looser over the years, they remained anchors for me. Sadly, they are both gone now.

Even though I was away at college, my mother expected to talk with me every day. I wasn't interested in such frequent communication. Dormitory residents could only get long-distance calls on one shared wall phone in each hallway. To my parents' chagrin, I asked them to limit telephoning me to Sunday mornings. I explained I knew they loved me and were always available if I wanted them, so I didn't need to talk every day. Looking back now as a grandma and knowing how much my children's daily pictures and texts enrich my life, I am sorry I restricted my parents' calls.

Dad wrote to me instead. I treasured these letters during college and even more after his sudden death in 1980. Here is an excerpt from one:

> Dear Miss Weiss,
>
> It is a pleasure to inform you that, on the basis of your actions as a daughter for the past sixteen plus years, you have been granted one million five hundred thirty-six thousand and ten advanced placement credits in the art of being a loving daughter. (This, by the way, is the maximum credit which can be granted.)
>
> We are also including the letter from MIT, granting you advanced placement credit in math. As you can see, they are quite a bit stingier than we are!

When I visited the MIT men's dormitories, there were no cuddly toys on display. I guess that was what Mom had expected.

However, she didn't realize that stuffed animals and dolls would be on almost every bed in my women's dorm. In my mind, this gave me license to rescue Margie and take her back to college. But that first Thanksgiving when I came home and looked for Margie where we had put her, she wasn't there. She wasn't anywhere. "Mom," I said with my eyes darting left and right, "Where is Margie? I can't find her in the armoire and she's not in my room."

"You mean that raggedy old thing?" Mom said with a sly smile after running her tongue over her teeth, "I didn't think you still wanted *that* so I threw it out."

I turned away from her. I knew Mom was trying to help me grow up. She thought I would never find a husband with Margie in tow. Part of me wondered if she might be right.

I began MIT as a math major, but in my sophomore year a problem-solving course stimulated my interest in biology. An amazing young professor, Charles Holt, designed and ran this introductory lab course. Sixteen years later, in 1982, I read an article in the MIT alumni magazine announcing that Dr. Holt had died suddenly at forty-five. The news that my favorite professor suffered a sudden heart attack while on vacation with his wife and four children shocked and saddened me. I think it was especially upsetting coming two years after my own father's death from what we thought was a heart attack. At the time I knew nothing about cardiomyopathy, so it never occurred to me that these deaths could have been anything but heart attacks caused by clogged arteries. Unlike a heart attack, an electrical disturbance that stops the heart from pumping is the cause of sudden cardiac arrest.

At the same time as I fell in love with biology through Professor Holt's course, I struggled with a linear algebra class. This prompted me to switch my major from math to biology.

CHAPTER EIGHT

Because of my continued interest in mathematics, I chose the subspecialty major biophysics and continued to take a math class every semester.

When I entered MIT, I planned to become a high school teacher, so I needed a college degree but not top marks. I aimed for a minimum of C's in my courses. But my competitive spirit and interest in the material earned me all A's and B's during my first two years.

By my junior year, I wanted to become a researcher and college professor. Then I obsessed over good grades to get into a top graduate school. I was struggling with a difficult biology course, Genetics, and feared I'd get a C in it. My friend Teresa was also in that class and had a similar worry. Since we had the same advisor, we made a joint appointment to talk with him.

When we arrived, we stood in the hallway outside Dr. Cyrus Levinthal's office because he wasn't ready for us. Equipment that we assumed wasn't being used lined the corridor. We searched for a place to sit but didn't find any chairs. Instead, we chose a top-loading incubator about two feet by three feet and two feet high, which seemed a workable substitute for a bench. We hoisted ourselves up and began to study while we waited.

When Dr. Levinthal came out to get us, he looked at us critically. "You shouldn't be sitting on expensive equipment."

We at once jumped down, taking our book bags with us. The corner of my lips turned down. "I'm sorry. I hope we didn't break it."

"You didn't damage it yet, but heavy objects resting on the door will eventually ruin the gasket." He lifted the cover of the incubator and showed us the rubber gasket.

Teresa's face was red. She brushed her wavy long brown hair out of her eyes and, looking at Dr. Leventhal said, "We didn't know it could be a problem. We won't do that again." This was not a good beginning.

Dr. Levinthal smiled and motioned for us to follow him. "Maybe I should put a few chairs in the hall," he mused. "Come sit down in my office."

Sitting in front of his desk, Teresa said, "We are having a challenging time understanding the material in the genetics class. We got C's on the first exam and are considering dropping the course before the six-week deadline for dropping a class. I'm worried because a C will hurt my chances of getting into a good medical school."

I twisted my hair and added, "The professors in the genetics class keep changing and they don't know what the previous lectures covered. Do I need Genetics for the biophysics subsection I'm in?"

Dr. Levinthal nodded. His raised lower lip caused his upper lip to protrude. "This is a new class. You are privileged to have Nobel Prize–winning lecturers that are excited to introduce you to their research. The problem may be that they don't know what to expect from undergraduates." Looking at Teresa, he continued, "You will have to retake Genetics if you drop it because you need it to get into medical school. I suggest you stick it out, but it's up to you."

Then he switched his gaze to me. "Requirements for the biophysics major are more flexible. But it's still essential that you master genetics." He rested his head on the palm of his hand before lifting it and putting his hands together. "I'll make you a deal. Continue attending lectures and taking exams after you drop the course and ask the professors to send me your unofficial course grade. If you get lower than a C, you will need to retake Genetics. Otherwise, you can do without it."

I leaned back in my chair and let out a big breath. "That sounds great. Thank you so much." With the pressure reduced, I enjoyed and appreciated the genetics lectures. I also asked more questions, which helped a lot. My course grade was B.

CHAPTER EIGHT

Genetics was the only class I ever dropped, and my transcript doesn't list it. I believe my struggle with this course fated me to teach it for thirty-four years and to become a geneticist. My experience made me a better teacher and more sympathetic to my students, many of whom had trouble with this difficult gateway course. I kept my personal battle with genetics a secret and only revealed it when I retired from teaching.

88

NINE

The Rabbit

"Actions speak louder than words." —Proverb

A poster on the MIT Biology Department bulletin board announced National Science Foundation summer internships at Yale University. Since I was a junior with no summer plans, I applied to the program. They admitted me and arranged for Dr. Jui H. Wang, a biochemist, to be my mentor. His laboratory was in the chemistry building, a short walk from Yale's International House. Although this co-op–style dormitory usually housed only foreign graduate students, they allowed me in because many of the international students were away for the summer. The house masters assigned me to a comfortable double room in the basement that kept cool during the scorching summer days.

My roommate hosted a farewell party the evening I arrived because she was leaving the next day. Even though we had just met, she invited me to her celebration. At nineteen I was already searching for a husband. Because shoes with heels created terrible pain in the ball of my feet, I fretted about whether to wear my low heels or flats. But, it was a casual backyard affair and my roommate recommended flats.

I met people from all over the world that evening, including one Alan Liebman, another Brooklyn Jew. Alan was a graduate student studying applied physics. When we stood next to each other in the yard, he was just a hair shorter than I. Since it was unusual for the girl in a couple to be taller than the boy, I was happy to be in flats. As we danced to the music playing on the radio, Alan looked at me with his medium-brown eyes. Moonlight lit up his smiling thick lower lip and slightly hooked nose. He was only twenty-three, but his wavy brown hair was already thinning. He recalls that I sat on his lap. I recall he telephoned the next day and we started seeing each other.

After the party, I had the room to myself for a couple of days before Lan, an ethnic Chinese student from Vietnam, moved in. We became fast friends and confidants. She gave me courage during that "long hot summer" of 1967 when widespread race riots and looting hit cities across America. When unrest came to New Haven, having only one small window in our basement room provided some relief for our fears.

We ate in the communal kitchen on the first floor, which had a bank of refrigerators, stoves, and ovens. Everyone shared these appliances, along with common pots, dishes, and silverware, to prepare our individual meals. We stacked our dishes in one of the dishwashers. If your plate was the last to fit in, it was your job to add soap and turn on the dishwasher. We took turns with other chores like emptying dishwashers or sweeping the kitchen floor.

In the biochemistry lab, I worked closely with one of my mentor's graduate students. She taught me how to use a spectrophotometer to assay the activity of the protein she studied. The overall question Dr. Wang's lab addressed was how living things can transform energy from their surroundings into a life force.

About a week after I met Alan, I woke up feeling sick and went to the Yale campus student health center. They took my temperature and sent me away because it was normal. I returned in the afternoon because I was worse, but I still didn't have a fever and they again told me to go away. That night I awakened at 2 a.m. and was very warm. Lan called the campus health center for me, and they sent a car to take me to the infirmary because of my high temperature.

The next day, besides being sick, I had hives all over my body and could barely open my eyes. Red itchy welts afflicted me throughout my childhood and adolescence. They often appeared on one side of my body, only to switch to the other side when I thought they were receding. These hives would cause my eyes to swell shut and my lips to puff out. They made me miserable. To prevent them, doctors told me to avoid many foods—including most fresh fruit.

I called both the lab and the International House to tell them I was in the nearby infirmary, anticipating that some people would come to visit. However, to my surprise, no one from either place showed up. Yet, amidst this unexpected solitude, Alan paid me a daily visit. Despite the hives making me look dreadful, he seemed unfazed by my appearance. In fact, he joked about bringing a paper bag next time, suggesting I could put it over my head to hide the hives if I wanted to. Beyond his humor, he brought me little presents during that week. I still cherish the brightly colored, cheerful enamel flower pin he gave me back then.

A few weeks later, when I entered Alan's apartment, the sweet smell of tobacco mixed with the appetizing odor of meat sauce hit me. Alan lived alone on the third floor of a large house. After

CHAPTER NINE

opening the door for me he ran back to the kitchen. "I'm making meatballs and spaghetti for dinner."

I found him confidently stirring, chopping, and frying. "It smells delicious. Can I help?"

"No need. I'm almost done." He handed me a sheet of paper. "Look at my weekly meal plan." I sat down at the table. His plan assigned a specific dish to every night of the week.

The table was set. Alan put the food on trivets with serving spoons. My stomach grumbled as I licked my lips. "Where did you learn to cook?"

Alan adjusted the flame under the tomato sauce to keep it at a slow simmer. "My mother taught me. I was an only child so I spent a lot of time with her." Although Alan's cooking was rudimentary, he was a more experienced chef than I—one of the many reasons I admired him.

As I attacked my dinner with a fork and knife, Alan's eyes opened wide. "You don't eat spaghetti and meatballs like that! Wind the spaghetti around your fork with the spoon." Alan tried to show me. "Watch how I do it." Perhaps because I am a lefty, I couldn't get it to work.

After repeated attempts I ground my teeth and squinted my eyes. My face was red. "Don't tell me how to eat," I yelled.

Then my anger evaporated and we both burst out laughing. To this day, whenever I get angry with Alan, I repeat, "Don't tell me how to eat."

After dinner Alan lit a fire. He was an expert at using paper and his stack of wood to get a fire going. There wasn't much heat in his apartment so this romantic gesture was actually necessary.

When Alan gave me a quick tour of the apartment I noticed a ragged lump of a sickly faded material on the top of his dresser leaning against the wall. It had an eye falling out that exposed a metal fastener. "What is that?"

"My Rabbit," he replied, "he's twenty-three years old and delicate. Don't touch him!" I smiled. Because of my experience with Margie, I understood Alan's relationship with Rabbit. Although he was an inanimate, silent observer, I was glad that Alan still had him.

Compared to my dormitory accommodations, I was amazed that Alan had a study, bedroom, kitchen, and living room with dining area all to himself. There was even a Steinway grand in the dining area! Alan did the decorating, house cleaning, laundry, and cooking himself. He painted a stunning brick red accent wall to surround the fireplace. He kept the place so orderly it reminded me of a luxury apartment pictured in a magazine.

After we returned to the living room in full view of the piano, I leaned forward in my armchair. "Do you play?"

"I'm taking lessons," he smiled. "My mother got rid of our piano when my baby brother, Kenneth, died from whooping cough. I was five then. This year I resolved to learn to play an instrument."

The room was warm and I took off my jacket. "What made you get a grand piano as a beginner?"

Alan threw another log on the fire. "I'm a student in a nearby music school and I used to pay $5 a week to practice on a piano there. Then the owner of the conservatory told me they had lots of pianos in storage and that if I paid the moving costs, I could rent a piano for the same $5 a week."

My mouth formed a circle as I lifted my head. "And they had a Steinway grand in storage?"

Alan nodded. "Yup. But it didn't fit on the narrow staircase leading up to my apartment. It was quite a show when they delivered it through a window. Practicing on it is a privilege."

This adventure had me spellbound. "Can you play something for me?" Alan worked the pedal with his foot and sang while he played Greensleeves. I gaped at him. "You have a great voice." I loved listening to him sing.

When Alan returned to the couch he fussed with his pipe cleaners, tobacco, and metal paraphernalia to clean and fill his pipe before smoking. "I started smoking a pipe when I worked at NASA in college. I learned about pipes from the foreign engineers there."

Although Alan looked distinguished and handsome with his pipe, I didn't like the smell and thought it was an unhealthy addiction. When he quit after a few years, I assumed the habit was gone forever. It disappointed me when he started smoking again five years later. The smoke was even more nauseating then. He continued smoking for over a decade before an oral cancer scare got him to quit again. This time I understood how hard it was for him to give up smoking and tried to remember the effort he was making. Now, I remain thankful thirty years later that he doesn't smoke.

Alan constantly made up puns and said silly things about our surroundings. This made it difficult to have a serious conversation, which both frustrated and challenged me. It took me years to fully appreciate his humor.

Although he often put his foot in his mouth with his jesting, everyone counted on Alan when they needed help. He transported large objects for friends in his station wagon. When others avoided greeting and chatting with people who looked different, Alan always welcomed those who were slighted. He saw everyone as a person first and their difference only secondarily. He was comfortable with homeless people, as well as people in wheelchairs or with Tourette's syndrome. Alan sometimes said the wrong things, but I loved that he did the right things. I had dated boys who talked about tolerance, inclusiveness, and feminism, but when it came to action, they never reached Alan's standard.

My nephew, Jeffrey Andrew Rothman, was born that same summer, in August 1967. He was my sister's first child. They lived in a small one-bedroom apartment in Brooklyn Heights, so they had to put the baby's crib in the living room. They signed up for a larger apartment nearby but the construction was very slow.

Diane took a leave of absence from teaching to stay home with the baby. Mom retired to help and took the Flatbush Avenue bus to see them every day. When she arrived, she instructed Diane to "rest" while she proudly took baby Jeffrey out in his carriage for walks. She also spent a lot of time gabbing with Diane. The intent was good, but I think there was too much help. With a daily maid and Mom's visits, Diane didn't have enough to do. She also missed her teaching career. This was a difficult adjustment.

Learning to cultivate girlfriends helped. One of her best friends gave her tips about cooking that Diane passed on to me. That is how I learned to rinse lettuce, dry it on paper towels, and then return it to the refrigerator to have a crisp salad. Through her friend, Diane got involved in the Women's League for Israel. They also attended a book club together. Only then did Diane realize she could read as well as anyone else and that she enjoyed discussing books, especially when they were relevant to her life.

Another of Diane's close girlfriends was also married to a lawyer, and her daughter was best friends with Diane's daughter, Karen. Yet another friend always advised on health foods and exercise. I got to know these women and their families during my visits to New York and through Diane's stories—they became like family. Diane continued to find it easy to make close girlfriends for the rest of her life. She confided in women upon first meeting them, and this often blossomed into friendship.

When I returned to MIT following my summer at Yale, I regularly took the New Haven Railroad between Boston and New Haven to visit Alan. The train was often hours late, repeatedly broke down en route, and lacked heat. On the weekends that I didn't go to Alan, he drove to see me.

CHAPTER NINE

When in New Haven, I stayed with Alan's international friends. When he came to MIT, he sometimes stayed in my friend Jerry (Gerald Jay) Sussman's room on a cot. I knew Jerry from high school. Junior year I set him up with my MIT "little sister," Julie Mazel. It was a good match and they were soon married. Jerry never left MIT or Julie. He is currently the Panasonic Professor of Electrical Engineering.

I saw a different side of Alan between visits. Unlike his in-person banter and silliness, he wrote long, thoughtful, and philosophical letters.

During Christmas break of my senior year, I stayed at school to finish a big paper. I told my parents and Alan, who would be in New York that week, that I couldn't spare the time to come home to Brooklyn. Imagine my surprise when Mom and Dad knocked at my dorm room door. They flew up and booked a room in a nearby hotel. Since I was too busy to come to them, they came to me! Mom at once sat down in front of my typewriter and asked what she could do to help. She typed up my entire handwritten report! They were still in the dorm on New Year's Eve when Alan called to wish me a Happy New Year. Instead of a lonely Christmas vacation, I had an unforgettable one.

Alan and I knew each other for less than a year when it was time for me to choose a graduate school. I applied to schools with top programs in biophysics, including Harvard and Yale. While I was interested in pursuing my relationship with Alan, we didn't have any formal commitment. Thus, I chose the slightly better known biophysics program at Harvard over Yale and remained in Cambridge with my friends for the time being.

My weekend New Haven railroad adventures continued. One morning after waiting for the train for hours while suffering severe menstrual cramps I called Alan to tell him I was giving up and going home. That afternoon he surprised me when he appeared on my doorstep. In response to my call, he jumped into his Country Squire Ford station wagon and drove up to collect me! I loved that secondhand car that Alan restored, with its leather seats, fake wood trim, and electronically controlled rear window.

Once when we went to Brooklyn together to visit our respective parents, Alan took me to a famous German restaurant. He warned me not to eat lunch so I would enjoy the expensive Luchow's dinner better. It took a while for them to seat us, so by the time the waiter brought the menus I was starving. The food listed was foreign and not at all appetizing (calves' brains). I was relieved when I found beef on the menu—something I recognized. But they described it as raw scraped beef. I told Alan that it looked good, but "Could we ask them to cook it?"

He explained they only scraped the beef while it was raw, then of course they cooked it. Alan used his best German to order our dinners. When mine came, the beef was raw! The waiter told us he usually warned patrons about this dish, but Alan's excellent German convinced him we knew what we were ordering.

During the winter break in my first year of graduate school, Alan and I were in New Haven and went to see the recently released version of *Romeo and Juliet* with Leonard Whiting and Olivia Hussey. This movie portrayed Romeo and Juliet as children, rather than adults, as in earlier interpretations. After the movie, Alan asked me to marry him, and I accepted. We had been edging toward this decision for a few months. We didn't want to tell our parents of our engagement on the telephone so we kept our news secret until the next time we saw them in person.

Then Alan spoke to my father in private, where he formally asked for permission to marry me. Dad didn't expect this kind of ritual. It flustered him and he wanted to know if Alan had asked me yet! After they straightened that out, they emerged together. Dad seemed shell-shocked as he announced, "Cyril, Alan just told me that Susan and he are engaged!"

"Congratulations! When are you going to announce your engagement?" Mom asked.

"We just did," I replied.

Within the next few weeks, my father engraved formal announcements in his print shop, and Mom sent them to the extended family and friends. We set a date for the end of the summer, a little over two years after we first met. My mother found the venue. It was to be in a Brooklyn synagogue on Sunday afternoon, August 17, 1969.

Mom tried to involve us in picking the menu, flowers, band, and color of the wedding, but we left it all in her hands. She even found my bridal gown. She borrowed three wedding dresses in my size from her friends' daughters. I tried on each of them and chose the one I liked the best. The woman whose dress I wore came to our wedding ceremony to watch her dress go down the aisle again. The only thing Alan and I did to prepare for the wedding was to pick "our song." We chose the Beatles' "When I'm Sixty-four."

While Mom fussed about the wedding, we arranged our schooling and jobs so that we would be in the same city. Alan was close to the end of his graduate work, while I was just beginning mine. Our goal was for Alan to finish his thesis by the end of the summer. Failing that, he would complete it in absentia while working during the day. Alan applied for positions in industry in the Boston area so I could continue at Harvard. Eventually he expanded his search to include other cities that also had biophysics graduate programs. He accepted a research and development position at Xerox, and I transferred to the Radiation Biology and Biophysics Department at the University of Rochester to work on a PhD.

I was sure I would have hives on my wedding day. Despite what doctors told me, I knew my hives were more related to what was going on in my life than to what I ate. I got hives on the first and last days of school, during finals week, during important examinations, and whenever my sister came home from college for a visit. Hives also struck when I had a fever or when I went to a dance or prom.

Alan and I enter our wedding reception. FAMILY PHOTO

CHAPTER NINE

I warned Alan not to expect a pretty bride. He laughed and said he didn't care what I looked like. He would marry me anyway. "Just keep your veil over your face until the hives go away."

I didn't have hives at the wedding. No swollen lips or eyelids. It is now over fifty years since that day, and no matter what I eat, I never had hives again. Being married removed enough stress from my life to end hives forever. Getting rid of them was the best wedding present ever!

The only concern I had about my marriage was that Alan might not complete his PhD and could then resent my getting one. But within the first few days of married life, I realized this wasn't a problem. Alan was proud of my accomplishments and would always help me be the best that I could be, whether or not he obtained his doctorate. Alan's job with Xerox was contingent on his getting a degree within a year. Although this was tough, he worked during the day, drafted his thesis at night, and met the deadline.

Alan was observant of kosher food rules. I had learned to love cheeseburgers from my college meal plan. Because eating meat and dairy together is not kosher, Alan wouldn't serve cheeseburgers. Still, one of our favorite dishes was meat and cheese lasagna. I questioned why we couldn't eat cheeseburgers but could eat the lasagna. Alan looked at me in stunned silence. He had somehow not noticed that this meal had both meat and dairy. He removed it from our allowed meal list, and I decided never to bring up inconsistencies again.

During our first year of marriage, Alan laid out goals for us: owning a house, having children, and saving, in our lifetime, a quarter of a million dollars. He suggested we were more likely to achieve happiness if we could define what we wanted and work toward it. I was on board with the goal of having children and liked the idea of owning a home and saving money to provide security.

We thought we were in control. But as my mother used to say, "Man plans, God laughs."

TEN

Her Memory Is a Blessing

"Celebrate *your* child." —My sister, Diane Weiss Rothman

In 1971, after living on Ocean Avenue for twenty years, my parents downsized into a new co-op in Brooklyn Heights just a short walk from my sister Diane's place. By then, Grandma Jennie was eighty-four and failing. After consulting with Mom's brother and his wife, they moved Grandma to an old-age home called Grace Plaza in Great Neck, New York. This facility was better than any in Brooklyn and was an easier drive for my aunt and uncle. My parents told Grandma to tell the relatives that she helped pick out the home to let her save face. Saving face was particularly important to Grandma.

I am glad my parents made this change, but I understand why Grandma didn't like it. My mother told Grandma it was necessary because she and Dad were now old themselves and needed to live near their daughter. She said Grandma accepted that. Indeed, my father died within a decade of the move.

It surprised me to learn that one of Grandma's worries about no longer living with my parents was that she feared she

CHAPTER TEN

wouldn't see me when I came home for visits. By then I was married and lived in Rochester, where I was in graduate school. I didn't realize I was so important to her. My parents drove to visit Grandma every weekend, and I came along whenever I was home. The highlight of every visit for Grandma was playing cards with Dad. She loved that. This gave me the idea that we should hire a high school student in the area to play cards with her during the week. My parents agreed and made this happen, enabling Grandma to enjoy card games every day. I feel good that I helped give her pleasure in those last years.

Diane gave birth to Karen Stephanie, with blond hair and blue eyes, on December 23, 1972. Karen and her father Ronnie shared the same birthday. The doctor induced labor because he feared the Christmas holiday traffic might prevent Diane from getting to the Manhattan hospital in time once her labor started. It was a short, easy birth. I flew in from Chicago, leaving Alan to attend a string of holiday parties on his own.

Mom and I transformed the blue crib that Jeffrey had used as a baby into one for a girl by painting it pink. While I was there, Ronnie didn't visit Diane and Karen in the hospital. Diane thought all the guests and commotion kept him away, so I returned to Skokie, the Chicago suburb where I lived. Ronnie was anxious about having two kids in their tiny apartment. With their son Jeffrey's bed in the living room, they had to put Karen's crib in the kitchen. Eventually, they couldn't wait any longer for the construction workers to complete the Clark Street building and moved into a nearby co-op at 75 Henry Street.

After Diane's four-year childcare leave was over, she taught a gifted first-grade class nearby in P.S. 29. Parents insisted the principal keep her in this coveted slot for many years because of her sterling reputation. Diane frequently told me how privileged she felt to repeatedly witness the miracle of a child learning to read. She also described her students' parents' worries that their children lacked ability in math or sports or were too shy. She told them to focus on their child's strengths instead of trying to make them fit another mold. Her counsel was "Celebrate *your* child." She reminded me to do the same for myself and my children just as she struggled to follow her own advice.

One of Diane's favorite activities was sharing stories from our childhood with her students. She would enchant them with tales of her chick's passing, the loss of my lovie doll Margie, and other special memories. I learned about this storytelling accidentally when I paid a visit to my sister's classroom. There, I met one of her former students, now a fifth grader, who had been sent by her teacher to deliver a message. The moment she realized she was face-to-face with the real "Susan," her excitement was palpable. She eagerly recounted her favorites of our stories and asked, "Did this really happen?" This interaction made the impact of my sister's storytelling very clear to me.

Inspired by the excitement these stories brought to Diane's students, we planned to create a children's book of Diane's memories and my illustrations. Alas, we never got to this. We thought we had forever . . .

Diane took four-year-old Karen to visit Grandma Jennie after Grandma fell and broke her hip. The break had confined Grandma to bed and would soon lead to pneumonia and death. Little Karen was puzzled and upset to see her great-grandmother helpless and all alone in a hospital bed. Karen stamped her foot and pointed her finger in anger. "Where is her mommy? Why isn't her mommy taking care of her?" By then, Grandma Jennie's mother had been dead for sixty-eight years! But it terrified little Karen that someone might have to deal with a serious illness

CHAPTER TEN

without her mommy. Little did she know that the best any of us can hope for is to outlive our parents. Heartbreakingly for Karen, she never had that privilege.

Years after Grandma Jennie died, my sister told me Grandma had skimmed cash off the weekly grocery expenses for years to use for pocket money. My parents knew about this at the time but decided not to humiliate Grandma by confronting her. Grandma didn't need this money, because my mother's brother Herb and his wife sent her an allowance for small purchases. But she put their money in a bank account to have an "estate" when she died. This was important for her pride. Grandma rarely bought anything for herself or others. The only physical thing I have from her is a round, stainless-steel pendant with a flat, dark purple rhinestone that my parents gave her toward the end of her life.

Grandma realized she had made mistakes. She appreciated her in-laws' commitment to her. Still, she thought it a sad commentary on her life that, "My in-laws care more for me than my own children."

Considering all the time I spent with Grandma Jennie, it's a shame that our relationship wasn't deeper. However, she was proud of me, and I was proud of her. She taught me a lot of things, including the concept of family illness. It was a blessing to know and to love one of my grandparents. We would name our daughter with a J name—Judith—after Grandma Jennie.

Meanwhile, Karen grew and learned about life. She was an adorable, sensitive child. One day about a year after Grandma Jennie died, Karen was talking with her father about an animal that died. He explained that all living things die. As Karen digested this, she became terribly upset. "You mean someday I'm going to die? No, no, no!!" she wailed. "I don't want to die!! No, no, no!!! I want to keep living!"

Not knowing how to console her, her father blurted out, "Don't worry, Karen, you won't be old enough to die for an awfully long time. By then, they will probably have a cure for death!"

Alas, there was no cure for death when Karen died thirty years later. The problem was, on the face of it, very minor. Just one mistaken letter in the six billion letters in her DNA. This included two versions of every gene, one from each parent. Even if we only consider the sixty million letters that code for proteins, one mistake in sixty million seems pretty good. Unfortunately, this one mistake reduced the level of a protein found in Karen's heart muscle and this caused her sudden death. For others, one mistake in six billion causes an increased risk for treatable cancer or other heart disease. Surely, we should be able to fix just one mistake.

There are medicines on the market that treat certain genetic mutations without gene therapy if patients and their doctors know that they have the mutation. For example, simple supplements of vitamins and minerals can treat some metabolic genetic diseases.[1] Administering an artificial or genetically engineered enzymes can treat mutations that inactivate an enzyme.[2] Other medicines inhibit the synthesis of toxic substances that accumulate as a result of mutation[3] or provide small molecules that stabilize a protein that misfolds because of a mutation.[4]

The modern form of humans evolved about two hundred thousand years ago, but gene therapy didn't successfully cure a disease until 1990, when it healed a four-year-old girl with a severely compromised immune system. She couldn't make any ADA protein because she inherited defective ADA genes from both her mother and father. To cure her, scientists inserted a good copy of the ADA gene into her DNA using a harmless virus engineered to code for ADA.[5]

Then, in 1999, there was a major setback in gene therapy when a procedure caused an immune reaction that killed a patient. In the following twenty-three years, the Food and Drug Administration approved about twenty gene therapies using better virus vectors.[6] One such therapy is for spinal muscular atrophy, which is caused by not having a good copy of the *SMA1* gene. In this treatment, a single infusion of working copies of the *SMA1* gene can supply good copies of the gene that keep working for the rest of the child's life. However, the infusion must take place in a young infant before the absence of *SMA1* damages their nerves.[7]

Mutations that reduce the level of a single protein are a major source of genetic disease. This was the case for the immunocompromised four-year-old girl. Likewise, the Ashkenazi mutation in my family causes heart disease because it reduces the level of the FLNC protein.[8] Several innovative technologies promise to revolutionize therapy for these types of mutations.

Novel RNA therapeutics are being developed to turn up the expression of the good copies of genes to make up for the loss of expression of mutant copies. Another idea uses the method championed by mRNA COVID vaccines. Researchers inject cells with mRNA that codes for and stimulates synthesis of the missing protein, rather than injecting mRNA coding for the COVID virus spike protein.[9]

Yet another approach is to use the DNA editing system, CRISPR [clustered regularly interspaced short palindromic repeats], which has immense potential to cure genetic disease.[10] Currently, several clinical trials use CRISPR's ability to target a specific sequence in a gene for destruction. For example, the adult hemoglobin red blood cell protein that moves oxygen and carbon dioxide around the body doesn't work properly in people with sickle cell anemia. CRISPR cures this by inactivating a gene that normally turns off the synthesis of fetal hemoglobin. Although usually only fetuses and not adults use fetal hemoglobin, it can substitute for compromised adult hemoglobin. In theory, CRISPR can also inactivate bad genes that make a "poison" protein. In the future, scientists may be able to use CRISPR technology to fix any DNA error where it occurs in the gene.

Currently, knowledge of most genetic mutations, including the one in my family, warns of an increased likelihood of disease. The best response is for the patient to seek repeated clinical examinations to allow for early disease treatment. This is important and can save and improve lives. However, gene therapy may soon allow us to take the added step of completely correcting mutational errors.

Currently, knowledge of most genetic mutations, including the one in my family, warns of an increased likelihood of disease. The best response is for the patient to seek repeated clinical examinations to allow for early disease treatment. This response can and can save and improve lives. However, gene therapy may soon allow us to take the added step of completely correcting mutational errors.

ELEVEN

Becoming a Geneticist

> "The 'awesome power of yeast genetics' has become legendary and is the envy of those who work with higher eukaryotes."
> —Dr. Fred Sherman, my graduate school mentor

As children, we assume our fathers are God-like and immortal. This is a tough myth to abandon even in the face of reason. We ignore obvious cracks in our parents' power, refusing, as long as possible, to grapple with the realization that they are mortal.

Alan and I were in our shared office in our Rochester home when the ringing phone broke my concentration. I tossed aside the yeast genetics paper I was reading and reached for the telephone receiver. Our rotary dial phone sat on my desk under the dark walnut wall shelves Alan had sanded to a high polish. From her first word, "Hi," I recognized Mom's voice.

We had agreed to speak on Sunday mornings, so it both annoyed and worried me to hear from her that Tuesday night. I looked at Alan smoking his pipe at his large, weathered, secondhand desk opposite his orderly bulletin board. "Hi, Mom. Is everything OK?"

CHAPTER ELEVEN

In my mind, I saw Mom holding the receiver to her ear just below her short bouffant haircut and running her tongue over her front teeth. "I know it's not Sunday, but you said I could call anytime if it was important." These words tumbled out in a single breath. "Important" to Mom rarely met the criteria of "important" to Alan and me. Mom longed to know what was happening in my life minute by minute. Left unregulated, she called several times a day. This strained my busy schedule; we settled on one weekly call on Sunday morning.

Realizing that Mom was on the phone, Alan stood to his full height of five feet, four inches. He leaned on his armchair and smiled as he slid his hand across his neck signaling that I should cut the conversation short. I stared past Alan at our framed diplomas and twisted the olive-green telephone cord. "Can it wait until Sunday, Mom? I'm busy studying for an exam on Friday."

"No. We need to plan a visit for this weekend. We have something important to tell you, and we don't want to do it over the phone."

I jerked my head up. This didn't sound right. Mom usually pumped me for information not the reverse. I decided we needed to see them. "OK, this weekend is probably fine. I'll check with Alan." Biting my lip, I realized that my parents' main talking points were details of my sister Diane's life. Was it something about Diane or the kids? Diane was on maternity leave caring for her one-year-old daughter, Karen. Her son, Jeffrey, attended first grade. "You have me worried. Can you tell me about it now?"

"It's nothing to worry about," Mom said in a calm, matter-of-fact voice. "But it's better if we discuss it in person."

Alan agreed to the visit, and we tried not to focus on it until Saturday. We bought a roast and straightened the house in preparation. Because Mom was always cold, we closed the windows, turned up the heat, opened the vents, and dressed in summer clothes before we left to pick my parents up from the airport in our fall jackets.

While driving, Alan tried to distract me with silly puns. I stared at falling red and gold leaves in their death throes. We met my parents at their gate. Dad was his usual cheerful self, but his tall frame traversed the terminal a tad more slowly than I recalled. During the ride home, Mom and Dad spoke about their new co-op in downtown Brooklyn and what a treat it was to live across the street from Diane and her kids.

When we opened our door, warm air with the odor of roast prime rib hit us. Mom complained, "Why is it so cold in your house?" I gave her a mug of steaming water to help warm her up because she didn't drink coffee or tea.

Dad winked and put his arm around Mom's slim waist in her red-and-white A-line dress and drew her to him to bestow a peck on her cheek. "Look at my beautiful hen!" Mom beamed and blushed even though she missed the days when Dad called her his chick. I walked across the linoleum floor in our large eat-in kitchen and opened the oven door to check on the roast. Dad called, "Where is my baby chick?" Embarrassed by the nickname, I rolled my eyes. Still, I reveled in being so obviously treasured.

When I returned from the kitchen, I passed the dining alcove with my parents' old table and chairs. I loved the delicate beige material with broad ribbon-like shiny magenta stripes that covered the chairs. In an earlier visit, Dad came out of the shower and put his wet towel down on one of these chairs. "Dad," I cried, "don't put a wet towel on our dining room chair!" Indeed, the towel caused a small water stain. But this was of no import compared to the hurt and disappointed look on my father's face. How could I chastise my wonderful father for putting a towel down on his old chair? Dad recovered quickly and never mentioned this incident. How I wished I could take back those words!

Mom and Dad were sitting on the living room sofa and looking at each other. Alan and I faced them on upholstered armless chairs covered with tufted Native American fabric that itched our bare thighs. The newly installed white spotlights over the

CHAPTER ELEVEN

sofa accentuated the fine lines on Mom's lovely face. During my childhood, while Mom was recovering from nephrosis, she had glassy eyes and a gaunt angular face and body. Now she was clear-eyed, round, and more beautiful. Red lipstick and nail polish were her only makeup. For years, Dad dyed Mom's premature gray over our bathroom sink. Since restoring her natural jet-black color was too harsh, they settled on deep brown. After Mom retired, she went to the beauty parlor every week. As a result, her hair and nails looked elegant as she sat there on our living room sofa.

From our seats opposite my parents and perpendicular to the living room picture window, we could see the large front yard and a lone cherry tree. Dad shielded his eyes as headlights from a car turning in front of the window lit his graying but still full head of hair. Although he wasn't heavy, I noticed a paunch about his belly for the first time.

Mom wrapped her legs in our sofa blanket and buttoned her cable-knit sweater, emphasizing her round shoulders; she wasn't young anymore. Mom's eyes took in my framed charcoal sketch of Alan before she returned her gaze to my father.

He adjusted his collar, sat up straight, and said, "We wanted you to know that I get chest pains from climbing stairs. It happens in the subway station going to work."

My throat tightened. "Did you have a heart attack?"

Dad took out a handkerchief and blew his nose loudly. "No. It wasn't a heart attack. I thought it was indigestion, but the pain kept coming back."

Alan paled and leaned forward. "What do your doctors say?" This was a surprising question from Alan, who never went to doctors.

Dad licked his lips. "The doc said I have angina pectoris."

I was breathing rapidly. "What is angina pectoris?"

Mom warmed her hands on the mug and sucked in air through her front teeth. "It's a restriction in the arteries that feed

Dad's heart," she said. Then, while looking straight at Dad she added, "He needs to stop running upstairs two at a time."

"You still vault subway stairs two at a time?"

"I'll stop that now." He continued sitting erect with unblinking eyes. "It isn't a heart attack. It isn't life-threatening. And it didn't damage my heart."

My breathing slowed. "How do you know it didn't damage your heart?"

Dad looked straight at me. "From an electrocardiogram. I have a normal heart rate and rhythm." I didn't know then that doctors would have needed to take an echocardiogram to see if the structure of Dad's heart was all right. Nowadays a heart disease diagnosis would suggest genetic testing; back in 1970 there was nothing more to do.

"Did you see a specialist?" My fears were easing.

"Yes. Mom's cousin, a cardiologist, recommended a doctor."

"What is the treatment and prognosis?"

Dad stretched out his arms and rolled his shoulders. "I can exercise and climb stairs one at a time." He glanced at Mom as he patted her knee and then faced me. "When the pain starts, I put a nitroglycerin tablet under my tongue. That opens the arteries and stops the problem." Reaching into his pocket Dad pulled out a small bottle. "I carry these pills all the time. My next appointment with the cardiologist is in six months." Dad adjusted his glasses and rubbed the bridge of his nose. "Also, I eat a healthier all-vegetable lunch at the automat instead of a beef sandwich and ice cream."

Mom's face was untroubled. "You needed to know not to worry if you see Dad put a pill under his tongue."

I cleared my throat and twisted my wedding ring. "Are you telling us everything?"

Dad smiled. "We've always been honest with you."

It was true. As far as I knew they were honest, especially about bad news. They didn't hide things to shield me, understanding

CHAPTER ELEVEN

that such deception would result in constant fear even when nothing was wrong. I flopped back in my chair, relieved.

Now that they had told us about the diagnosis, my parents steered the conversation away from illness. Dad engaged Alan in politics and asked me to explain the discovery of recombinant DNA that he read about in the *Times*. My parents also gave us the latest on Diane and the kids.

My father was fifty-seven. He had good medical advice and a sensible treatment plan. Evidence suggested he could live a long time. Thus, at twenty-three, I returned my attention to completing my graduate studies, furnishing our home, and planning for babies after graduation. While I didn't focus on Dad's condition, neither could I ignore this crack in his immortality.

We had come to Rochester two years earlier when Alan began working at Xerox and I transferred to the University of Rochester. With a master's from Harvard, I was ready to start on my doctoral thesis. First, I needed an advisor. I looked for a top scientist in the department whose research interested me. I considered working in X-ray crystallography, but eventually I wasn't convinced that the supervisor would be able to give me enough of his time.

Meanwhile, another distinguished faculty member, named Fred Sherman, gave a seminar that caught my interest. By then the Nobel Laureate Marshall Nirenberg, whom I had met on my Westinghouse trip to Washington, D.C., had cracked the genetic code. He used molecules in a test tube to show that the code consisted of ordered groups of three of the four possible DNA letters. There are sixty-four such unique combinations, called codons. Sixty-one codons specify one of twenty protein building blocks (amino acids) while the remaining three codons are used as stop signals. Later, another scientist, Charles Yanofsky, confirmed that bacterial cells actually used this redundant genetic code. The same codons, whether in a test tube or in living bacteria, coded for the same protein building blocks.

Dr. Sherman's talk reminded me that bacteria lack a nucleus and differ vastly from higher cells like ours. Sherman used bakers' yeast, which have nuclei, to show that higher cells use the same genetic code as bacteria. It wasn't possible to sequence DNA in those days, so Dr. Sherman deduced changes in the DNA from the changes in the sequence of the amino acids in the protein it encoded. He found that the same codons corresponded to the same amino acids in bacteria and yeast. It seemed the DNA code was universal. This work enthralled me, and I asked to join his laboratory.

Dr. Sherman agreed to take me on as a student, but since genetics wasn't on my transcript, he wanted me to listen to the undergraduate genetics course taught that semester. In the beginning, Dr. Sherman spent hours each day teaching me about yeast in his tiny office packed with books and papers.

His boyish face and mismatched socks belied his brilliance. "We use the simple single-cell organism baker's yeast as a model for human cells," he said. "Yeast are easy to grow and experiment with. I'm excited because I think what we learn about the basic biology of yeast will also be true of human cells. I can't believe they pay me to have so much fun!" His prophecy proved to be correct. Together with his longtime colleague Jerry Fink, he taught a Yeast Genetics and Molecular Biology summer course at Cold Spring Harbor for seventeen years. This course had a significant impact on creating a productive yeast genetics community. Part of the community's phenomenal success was that it continued Dr. Sherman's example of sharing ideas, reagents, and strains. Course graduates include three Nobel Laureates and twenty-eight members of the National Academy of Sciences.

Wearing an old suit and narrow tie, Dr. Sherman took out a pen and paper to illustrate as he spoke. "Yeast cells reproduce by budding off identical copies of themselves. They have two mating types, *MAT*a and *MAT*α."

What I learned in genetics was coming back to me and was clearer now. I twisted in my stenographer's chair. "Is this mitosis?"

Looking at me and nodding, Dr. Sherman continued, "Exactly, but they can also mate and combine their genetic material like a human egg and sperm. To mate them you use sterile toothpicks and mix equal amounts of cells of the opposite mating type on nutrient agar in a petri dish. If you look at these cells under the microscope two hours later, you will see they elongate toward a potential mate. We call these funny-shaped cells 'shmoos' after the creatures in Al Capp's Li'l Abner cartoon. In five to eight hours, the shmoos and their nuclei fuse with a mating partner to form zygotes. The zygotes bud off cells with nuclei containing all the chromosomes from both parents."

Dr. Sherman spoke for so long that the sun was setting and we had to turn on the office light. "Each chromosome encodes many genes and each gene spells out the code for the synthesis of a specific protein," he said. "The order of the codons in each gene directs the cell machinery to join the twenty protein building blocks—amino acids—in the correct linear order to make the encoded protein. The codons that correspond to an amino acid are called 'sense' codons. The final codon in a gene instructs the machinery to release the completed protein. These are the 'stop' or 'nonsense' codons."

Now we were getting to the crux of the matter. "One idea for your thesis is to use my iso-1-cytochrome c gene/protein system in yeast to isolate mutations that prevent the cell from recognizing nonsense codons that say the protein is complete. Such mutations should help us find the machinery used to terminate the synthesis of proteins."

When I entered graduate school in biophysics, I expected to join my two loves, biology and mathematics. "Is this biophysics? It doesn't seem to have any physics in it."

Dr. Sherman chuckled. "Biophysics is a misnomer. When they chose the name, they assumed that the many physicists turned biologists who joined the field would employ physics in their biology studies. Instead, these former physicists developed molecular biology and genetics. These luminaries included Francis Crick, the co-discoverer of the structure of DNA; Max Delbrück, co-founder of phage genetics; Maurice Wilkins, who contributed to the discovery of DNA structure; and Seymour Benzer, who developed the genetics of the fruit fly and used it to connect genes to behavior."

Whether it was biophysics or molecular biology, I was excited about the science Dr. Sherman described, and "Nonsense Suppression in Yeast" became my thesis project.

I sat on a lab stool next to my three-foot-high bench silently counting the number of red vs. white yeast colonies on my petri dishes. "Twenty-four, twenty-five, twenty-six, twenty-seven . . ." The lab smelled like a bakery with the sweet odor of yeast. My spine ached as I leaned over the plates, because there was no backrest on my stool. Behind me one of Dr. Sherman's technicians, Maria, sat at her bench with her radio playing.

Kathy, another tech, burst into the lab and plopped onto a low lab bench against the wall near Maria. She swung her legs as she gestured with her arms and smiled. "My columns won't need me again for fifteen minutes. John's kindergarten teacher gave me a great report at open school night. John is doing so much better now. Patrick was supposed to take him this weekend. Naturally, he didn't show up."

Maria turned off her radio. "How did Tim react to having Johnny for the weekend?"

Kathy tossed her long brown hair and smiled. "Tim was great as usual." Kathy's second marriage, to Tim, was a happy one.

CHAPTER ELEVEN

Maria turned in her lab chair to face Kathy. "Can you come to Bible study this week?"

"Will it be at the Parkminster Presbyterian Church again?"

"Yes. I can drive. I'm appointed to chair their fundraiser. We're organizing a huge rummage sale."

I sighed as I realized I had lost count of the colonies and had to start over again. I decided to wait until Kathy left us.

Dr. Sherman stuck his head into the lab, announcing lunch. I dropped by my shared office space to snatch my purse. We picked up Patricia, a postdoc in the lab, and headed to the cafeteria, where Chris Lawrence and Mike Resnick met us. Chris and Mike were professors who also worked with yeast. I was part of this lunch clutch where we discussed research until I left Rochester five years later. Dr. Sherman made these interactions enjoyable. He was a Columbo-type character often asking seemingly naive questions and then coming up with profound suggestions. We also discussed life and got to know each other. Dr. Sherman told me he was born in Minneapolis to Jewish Ukrainian immigrant parents. "They didn't know much about names so my birth certificate reads 'Freddie,'" he said. "We lived in a few rooms behind my father's grocery store. I thought I was one of the richest kids in the world because we always had enough to eat." Dr. Sherman used humor to make people feel included and comfortable. For example, he would break the ice with a lonely graduate student at a meeting by asking "How are you doing?" The surprised student would often respond "I'm fine, how are you?" to which Dr. Sherman would say, "Well, I think I'm fantastic—but not everyone agrees with me." He could bring out a smile in even the most desperate situations. As he lay dying, connected to tubes and monitors, his nurse told him she had to go to get something and would return in a minute. He responded, "If you don't mind, I'll just wait for you here."

After lunch, I went back to the lab to complete counting my colonies. Maria was on the phone talking about her rummage sale. We didn't get along very well, but I had an idea to improve the situation.

I stopped by her bench. "Maria, I overheard about the rummage sale you are organizing. Alan and I are replacing our secondhand living room furniture. Could you use it for your church sale?"

Maria jumped up and clapped her hands. "Could I ever!" Her response made me feel great. This interaction transformed our relationship for the better.

A few years later, Patricia, the postdoc who ate lunch with us, had to deal with a family problem. Her father suffered a heart attack when visiting. He was too ill to travel home to Brooklyn and so remained in Rochester to recuperate. I felt sorry that Patricia had to arrange for his care and wondered how she would keep her work going at the same time. I was thankful I didn't have that burden. Only after my father died and I was jealous that Patricia had the privilege of helping her father recover did I realize how stupid and selfish my thinking about a burden was.

Upon completion of doctoral graduate studies, it is traditional for biologists interested in an academic career to switch universities and continue their training in a new lab as a postdoctoral student. However, because of Alan's job at Xerox, it was difficult for us to leave Rochester, so I remained in Dr. Sherman's lab for my postdoctoral work. After two years of trying to conceive, I finally got pregnant during my last year as a graduate student, and our son was born when I was a postdoc.

Our baby was small, under five pounds. Except for having trouble keeping down food, thankfully he was healthy. We didn't know then that he was at risk for sudden death. However, we knew that young children could die. Besides my uncle Eugene's death at four, my husband's eighteen-month-old younger brother,

CHAPTER ELEVEN

Kenneth, died when Alan was five. He remembers his mother banning music from their house for years afterward because she was still grieving. She still talked about the loss when I met her after Kenneth had been dead for over twenty years.

While we were in the hospital, we had to decide on a name for our baby. This became complicated after my mother came to visit. "I think you should know," Mom said, "Alan's mother called me. She said that if you don't name the baby after Kenneth, she won't have anything more to do with you. She's tired of waiting for someone to name a child after Kenny and was upset and disappointed that none of the cousins did it. Now she feels Alan must do it!" It is customary for Ashkenazi Jews to name their children after a deceased relative. The purpose is to both keep the memory of the relative alive and instill the relative's admirable qualities into the child. Some believe the souls of the departed live on through their namesakes. My parents followed this tradition (to match first letters) when they named my sister Diane Elaine after Dad's father David and Mom's grandmother Elke. They named me Susan after Mom's father Simon.

But Alan didn't want to name our baby after his brother. "Who wants to name a baby after someone who died so young? It's bad luck. Also, my whole life revolved around this loss. Now I want to move on."

My father suggested we name the baby after two people: Kenneth, who died young, along with someone else with a normal life span. He proposed we choose his mother Marion, who lived to have children and grandchildren. Alan agreed to this. The full name we gave our baby was Michael Kevin. This appeased Alan and thrilled his mother.

Alan's parents were uncomfortable that we planned to leave Michael with a babysitter on weekdays. Before I returned to work, they announced, "Taking care of a baby is not something we planned to do in our old age. However, if you're set on working, we will raise him for you." This comment flabbergasted me. I had no interest in giving up my baby!

"Thank you, Mom and Dad, for this generous offer," I said. "However, we want Michael here with us for all the hours and days that we are not working." Alan shared in all the work as the chores mounted. He changed diapers and did housework. He still cooks dinner six nights a week and does the grocery shopping, laundry, and vacuuming.

We found a great family to take care of Michael when both of us were at work. Margret, who had a one-year-old girl of her own, was happy to take him in. We warned her that Michael spat up all the time. She was unperturbed. Armed in old clothes and burping cloths, she was ready for him. We installed a crib in her house and brought diapers, bottles, formula, and changes of clothes each day. This family continued to provide Michael's daycare until we left Rochester when he was two.

At nine months, Michael was a speedy, crawling demon. But he was still constantly spitting up. Rather than put him in a playpen or have him leave a trail of his vomit all over our carpet, I crawled with him holding a cloth diaper under his mouth to catch the mess. This astounded Alan's parents. "Look at the attention he is getting!" they said gleefully. After that, there was no more talk about them raising him for us.

A few years later, three-year-old Michael had a fever and couldn't go to daycare. We called my mother for help even though she lived far away. She jumped on a shuttle flight from New York to Chicago to take care of him so we could go to work. During that visit she spoke to me about that summer day in Free Acres, decades earlier, when she went to the hospital. She confided, "You were such an independent, busy child, and never had time for me. I didn't think you cared about me at all until Dad told you I had to go to the hospital. Watching you cry that evening made me realize you did love me. Despite being desperately ill, that was cause for me to celebrate."

TWELVE

Professor Sue

"That happened to me once." —Alan Liebman

By 1976, it was time for me to start my own laboratory group. Alan was also ready for a change from Xerox. Although Alan put feminists down with his comments and jokes, he automatically adjusted his job search to help me get a suitable position. In contrast, my friends' husbands talked about feminism but did nothing to accommodate their wives' careers.

To get jobs in the same city, we both applied in the Boston, New York City, and Chicago areas. I started my search first since academic positions had a longer lead time. Clark University near Boston, Brooklyn Polytech, and the University of Illinois at Chicago (UIC) each invited me for an interview.

My advisor, Fred Sherman, warned me not to be condescending during these visits. He said young people commonly make this mistake on the job trail. This suggestion surprised me because, although I certainly had many faults, condescension wasn't one of them.

CHAPTER TWELVE

I saw what he meant, though, during my first interview when I visited Clark University. Their equipment and staff were so much beneath what I had experienced at MIT, Harvard, and the University of Rochester that it was hard not to be condescending when they asked, "What do you think of our facilities?"

I got offers from all three departments. That left it up to Alan to find a job in a matching city. I negotiated with each of the schools to let me delay my response until Alan completed his search. This posed no problem for the University of Illinois and Brooklyn Polytech. In contrast, the head of the biology department at Clark University responded angrily. "You women and Blacks, who do you think you are? Any white man would jump at this offer." After that, we crossed Clark off our list.

I liked that I could walk from Brooklyn Polytech to my parents' and sister's homes in Brooklyn Heights, but Alan thought it might be too close to the family. We picked the University of Illinois at Chicago because Alan received a matching job offer from Teletype in Skokie, a Chicago suburb.

I loved our two-story house on a tree-lined street in Skokie with large immigrant and Jewish populations. Most of the homes were thirty years old, but across the street stood a huge modern house that foretold what would happen to the neighborhood. The only challenge was that nearly all the women in the community were stay-at-home housewives, which hampered my ability to make friends. Still, UIC was a wonderful choice. The biology department welcomed me enthusiastically.

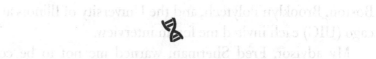

I stood in the lecture hall where my genetics class was about to start. My heart pounded. One hundred and seventy pairs of eyes stared at me. I had forgotten to bring my notes. I had no idea what to say. The bell sounded, compounding my panic.

That's when I woke up. Versions of this nightmare plagued me for decades. That morning I dressed quickly, grabbed my blue backpack stuffed with lecture notes, reference books, and course handouts, and drove to my office. There, I practiced my talk for the last time before walking to the lecture hall for my 10 a.m. class.

The passing period began at 9:50 a.m. Eager to prepare the blackboard and podium, I entered the lecture hall as soon as students from the earlier class started to emerge. The professor from the previous class was packing up to leave. Light bounced off his glasses as he lifted his head, displaying his distinguished gray beard. He stretched his arm out and signaled to me to stop.

"Wait outside with the other students until the room is empty." Having taken care of me, he returned to squeezing papers into his brown leather briefcase.

My shoulders slumped and my head drooped. I was thirty. Did he think I was still in college? I lifted my head and looked straight at him. "Thanks for the compliment, but I'm not a student. I want to prepare the boards for my genetics class that begins in a few minutes."

"Sorry for the mistake. I'm still not used to women professors. And your backpack fooled me. I'll help you erase the boards."

He wasn't so bad. "Thanks. This is my first lecture and I'm nervous."

After I wrote my notes on the board, I clapped my hands to get rid of the chalk dust. The students chatted as they entered the room and found seats but stopped talking and looked up when the bell rang.

I asked two students in the front row to help distribute handouts as I discussed the mechanics of the course, including turning in homework, quizzes, exams, and office hours. During my lecture, I wrote a lot on the blackboard to prevent myself from talking too fast. "I know you have all learned about mitosis and meiosis in high school and introductory biology. But please pay close attention as I review this again because we will examine it from a geneticist's point of view."

I pointed to my sketch of three pairs of meiotic chromosomes on the board. "Each chromosome has two sister chromatids that are held together by a centromere. In each pair the filled-in centromeres were inherited from one parent, while the open centromeres were inherited from the other parent. As you know, in meiosis, microtubules attach to centromeres to randomly pull the paired chromosomes to opposite poles in the cell. What is the chance that the microtubules will pull all three chromosomes

Cartoon of meiosis. Centromeres (circles) hold sister chromatids together. Open centromeres came from one parent, closed centromeres from the other parent. AUTHOR CREATED

with the filled in centromeres in the same direction?" Silence. I waited for what seemed like an eternity but was really only two minutes and asked again. "What is the chance that the centromeres of these three chromosomes inherited from the same parent will go together in the same direction?" More silence. Students looked down and wrote in their notebooks. Then a few tentative hands appeared. I pointed to the first student to raise her hand. "Yes, the student in the third row, please stand and give your answer."

Her notebook tumbled to the floor as she stood. "The chance is one-quarter."

I nodded and smiled. "Thank you very much. One-quarter is correct." In no time, the bell rang signaling the end of class.

The students stayed to copy down the last few notes before heading out. Some crowded around me, vying for my attention as they asked questions. I loved this.

I survived my first class. There was a spring in my step as I walked in the sunshine back to my laboratory.

In addition to teaching, research was an important part of my career. During the academic year, I pursued my research while teaching undergraduate and graduate courses. Summers were pure research.

In the early days, my lab studied mutations I isolated as a graduate student that inhibited cells from recognizing the ends of genes. We looked for other mutations that enabled these mutant cells to regain their lost ability. The second tier of mutations pinpointed yet other cellular components involved in recognizing the ends of genes. This kind of back and forth is a common genetic research ploy. We first isolate mutations that cause a problem in a process under study and then isolate other mutations that reduce the problem. The researcher can then decipher which cellular components the mutant genes encode and how they affect the process of interest.

While the university funded my academic year's salary, it didn't fund my summer salary or incomes for my graduate students, postdocs, and technician. I needed money from outside grants to pay for these and for the supplies and equipment for my research.

Getting a grant is extremely competitive. When I was an assistant professor, the body of a National Institutes of Health (NIH) grant proposal was twenty-five single-spaced pages. Together with the Summary, Specific Aims, References, Budget, CV, and other form pages, a typical proposal was about fifty pages.

I worked diligently for four to eight months on each proposal and completed a rough draft a month before the deadline. This left me time to make changes after colleagues read and critiqued my efforts. I once had a proposal rejected five times before the U.S. Army finally funded it. As is true for other things in life, except for a bruised ego and emotional letdown, the number of rejections is of no importance. All that counted was that I eventually got the money to do the research.

My success in getting grants enabled me to support my research and brought recognition and money to the department and university. I never assumed I would get a proposal funded: It was always a surprise, great relief, and a delight when it happened. I was on a merry-go-round of grant writing because, after two to four years, funded grants had to undergo a competitive renewal. To ensure some financial stability for the lab, I maintained two or three staggered grants from either NIH, the National Science Foundation, the American Cancer Society, the Alzheimer's Association, or in later years the U.S. Army. Even if one grant renewal wasn't approved, another grant allowed me to continue paying salaries to my graduate students, postdocs, and technician.

I was determined to have both a rich family life and a fulfilling career. This required a bit of pioneering. Since it took several years for me to conceive our first child, we tried for another baby about a year after we arrived in Skokie. Having learned how to calculate my ovulation cycle the last time, I now got pregnant in the first month! We didn't want to announce it right away but thought my terrible morning sickness would give us away. However, nobody noticed my months of lingering ills because many other people were sick with the flu.

It was my turn to teach genetics in the fall of 1978 just a few weeks after Judy was born. It would be many years before the university arranged for formal maternity leave. Luckily, I had wonderful colleagues who wanted me to succeed. Dr. Eliot Spiess, a distinguished senior faculty member who studied mating in fruit flies, offered to teach my section of Genetics along with his in the fall quarter following Judy's birth. I returned the favor the next year when I taught his section along with mine.

Instead of taking Judy to a daycare home, as we had done for Michael in Rochester, we hired a nanny to come to our house and also help care for Michael. I had total confidence in her, so I could focus fully on my work when I was at school. Likewise, I forgot about my work entirely when I was home with the kids. I think the break from home made me a better mother, and the break from work helped enhance my creativity at work. The kids and family life certainly helped me put my work woes into perspective.

Despite overall support from my husband, there were some rough spots. One difficulty was my failure to get the empathy I craved as I dealt with problems at work and elsewhere. The trouble was, as soon as I told Alan about a problem, he set about trying to solve it for me. I wanted him to listen and sympathize without telling me what to do. He couldn't understand this and thought something was wrong with me. Many of these encounters took place in the wood-paneled office we shared in our Skokie home, where a small television set sat on top of a file cabinet. Alan used that TV to silently watch the stock ticker. One day, by accident, he came across an interview in progress with Debra Tannen about her new book, *You Just Don't Understand: Women and Men in Conversation*. This program caught our attention. We watched for about ten minutes, and that was enough to change our lives.

"Men and women see things differently," Tannen explained. "Men want to solve problems; women want sympathy and understanding." She noted that women, talking to each other about their problems, often respond with, "I know how you feel. Something similar happened to me." In contrast, men try to one-up each other by proposing solutions. At the end of the ten minutes, the interview convinced Alan there was nothing weird about me except that I was a woman. And I understood that his difficulty in responding the way I wanted him to was typical for men rather than a unique flaw in him.

After that, we each learned to reply to stories of difficulty with, "That happened to me once!" Even though it is a canned response, we find it comforting. Sometimes it is downright funny. Suffering from terrible menstrual cramps, I once complained of the pain to Alan. "That happened to me once," was his lightning-fast reply!

For over fifty years, Alan has supplied the emotional and practical support I needed to enjoy family life and become a successful scientist. In all this time, Rabbit sat on Alan's dresser faithfully watching over us. He is sitting there still.

THIRTEEN

Interlude: Pros and Cons of Genetic Testing and Screening

"Knowledge is power." —Thomas Jefferson

During the thirty-five years that I taught genetics, the field changed and grew tremendously. In 1996, after I had been teaching for twenty years, Alice Wexler published a powerful book called *Mapping Fate*. This book described the devastation her family suffered from Huntington's disease, and how her sister Nancy spearheaded genetic research leading to the identification of the gene responsible. I recommended *Mapping Fate: A Memoir of Family, Risk, and Genetic Research* to my Introduction to Genetics students because it illustrated the personal aspect of genetic research, the practical impact of genetic discovery, and dilemmas surrounding genetic testing.

Nancy Wexler was twenty-two in 1968 when her father told her and her older sister, Alice, that their mother had a disabling fatal illness called Huntington's disease (HD). He also revealed that there was a 50 percent chance that each of them inherited a dominant mutation that would cause them to develop the disease. Despite this devastating news, the sisters went on

CHAPTER THIRTEEN

to live remarkable lives that included changing the course of healthcare worldwide. I had no idea during the years when I assigned this book to my students how important the Wexler's story would be for my family.

Doctors can do little to treat or prevent HD, and it is always fatal. HD causes nerve cells in the brain to break down over time, leading to movement, reasoning, and psychiatric symptoms. Most HD mutations are fully penetrant, so everyone who inherits the mutation will get HD, although there is a range of ages at which the disease appears. Only recently has the Food and Drug Administration approved a drug, tetrabenazine, that helps treat symptoms.

Nancy and her father dedicated their lives to figuring out where the gene that causes HD is in the human genome and what the gene encodes. Alice is a writer and described her family story and their extraordinary contribution to science and medicine in her book.

In 1983, Nancy and her collaborators published a landmark paper in which they described how they mapped a disease-causing gene, *HTT*, to a specific chromosomal location for the first time.[1] To do this, they used an idea that David Botstein proposed, stemming from work in the model organism yeast.[2] Yeast was the first organism with a nucleus to have its entire DNA sequenced. The sequences of many different yeast strains showed that there were a lot of single-letter variations in the sequences throughout the genome. These little changes didn't cause changes in cells, but they could be detected by scientists and serve as markers for mapping genes. Dr. Botstein proposed that this would be true in human DNA, too, and it was. For the study to be successful, Nancy needed a large number of blood samples from families with HD. With determination, she identified and engaged an extended Venezuelan family grappling with HD and convinced them to donate blood samples for the study.

INTERLUDE: PROS AND CONS OF GENETIC TESTING AND SCREENING

This allowed her and her collaborators to map the location of the disease mutation relative to sequence variations in the genome. Nancy's tenacity, scientific excellence, and charismatic personality led to success. Her father, a prominent Wall Street financier, raised money for the project.

To use the mapping information to find the HD gene as rapidly as possible, Nancy convinced top scientists to work together rather than compete. Within a decade, the Huntington disease collaborative group cloned and sequenced the HD gene.

Learning the sequence of the gene revealed a great deal about the cause of the disease. Although there is still no cure, we now know that in people with the disease a short piece of the HD protein is repeated many times. It is analogous to a short story with the phrase "No, no, no, no, no!" where a printer's error mistakenly repeats this phrase in tandem six or more times. This expansion error causes the protein to fold incorrectly, causing copies of the protein to stick together and make an aggregate. We now know that the age of onset of the disease and the speed with which the protein aggregates is proportional to the size of the expanded repeat. I have used yeast to study this and other types of disease proteins that misfold and aggregate.

Once they found the disease mutation, anyone, including Nancy and Alice, could take a genetic test to find out if they had it. Such decisions are personal. The reasons to choose or decline testing can change. The Wexler sisters already knew they had a 50 percent chance of inheriting the mutation and contracting HD. For them, if they didn't inherit the mutation, taking the test could have removed unnecessary fear of HD from their lives. However if they had the mutation, the test would predict with certainty that they would have to face the devastating, incurable disease they watched their mother die of.

Both sisters decided against taking the test. For them, living with ambiguity seemed the best choice. As a corollary, they each decided not to have children. Nancy publicly announced that she had HD in March 2020, when she was seventy-five. Alice at eighty shows no symptoms.

About 97 percent of people who get HD have a strong family history of the disease. For the remaining 3 percent, some have a new repeat expansion or inherited a small repeat that doesn't always cause disease. Others may have a faulty family history. Genetic screening of the general population for HD would be of limited value since there is no available treatment. The only use would be to warn those with the mutation that if they had children, they would have a 50 percent chance of inheriting it. Options would then include ignoring the information, not having children, adopting children, or using preimplantation genetic diagnosis (PGD) to ensure that children didn't inherit the mutation.

PGD is a new procedure that allows genetic testing after in vitro fertilization when the embryo is at day three of development, at the six- to eight-cell stage. After testing, doctors implant an embryo lacking the mutation into the woman's uterus. For many, this is more acceptable than aborting a pregnancy based on a prenatal genetic test. Couples can use PGD to make sure that their children do not inherit diseases caused by a single known mutation in the family.

There is nothing doctors can do to delay or treat HD. However, there are other genetic diseases that can be treated or prevented and are, therefore, called "actionable." The American College of Medical Genetics and Genomics (ACMG) has released a list of eighty-one actionable genes.[3] They recommend that doctors inform patients if mutations in these genes are found incidentally (as secondary findings) when clinical sequencing is done for another purpose. The ACMG adds genes to this list every year following medical advances. Another resource dedicated to understanding the clinical effects of mutations is ClinGen, funded by the National Institutes of Health (NIH). As of January 2024, ClinGen's website categorized 140 genes as moderately, strongly, or definitively actionable.[4] Diseases caused by a mutation in a single gene are called monogenic. Genes on the actionable list are monogenic.

Illnesses that run in families but require the simultaneous presence of mutations in more than one gene are called "polygenic." Examples of polygenic diseases include Parkinson's disease, Alzheimer's disease, schizophrenia, coronary artery disease, and diabetes. Many mutations work together to cause each of these illnesses, and most of the genes involved are unknown. Nonetheless, researchers are developing ways to test for genetic risk. By comparing the frequencies of a large number of common variations in the sequences of genomic DNA in people with or without a particular disease, researchers can compile a polygenic risk score. When this enters clinical use, it will help people determine their personal risk.[5]

The pros and cons of genetic testing vary depending on the gene and disease in question. Learning of enhanced risk for a disabling and fatal disease for which there is no cure or effective treatment, such as HD, is of limited value. However, learning of enhanced risk for a disease that doctors can treat to improve the quality and length of life, like the cardiomyopathy in my family, is of profound merit. Unlike the HD gene, which always causes fatal disease, most mutations in actionable genes do not always cause disease. Thus, even if you learn you have a disease mutation, you may never exhibit symptoms. The information is a warning to inform relatives and to continue to follow up with clinical screening that can lead to treatment and a better life.

Some people prefer not to know, fearing that learning of a genetic defect will cause them to limit participation in life. However, not knowing of a genetic risk doesn't make it go away.

Patients with a strong family history, like the Wexler sisters, don't need a genetic test to know they are at elevated risk of having a mutation that can cause disease. However, for other patients, there is no warning from family history. For them, preemptive genetic screening could give lifesaving information.

Indeed, inheriting a bad gene from our parents is not the only cause of genetic disease. Several novel mutations arise in every baby. Generally, these don't lead to illness, but sometimes, as happened to my daughter, Judy, they do. We had no inkling of her genetic disease until she was thirty-two and critically ill with pancreatitis. While she was a patient at Stanford University Hospital, an intern noticed that her blood cells were too round. From this chance observation and follow-up blood work, he diagnosed her with a red blood cell disorder, hereditary spherocytosis.

This explained many of Judy's lifelong health issues, including high bilirubin, degeneration of her gallbladder, digestive troubles, stomach pain, susceptibility to infectious disease, and pancreatitis. The cause was a dominant mutation in the *SLC4A1* gene that neither I nor my husband have. If Judy had been aware of her mutation earlier, she could have used folic acid and B-vitamin therapy to avoid years of hardship. Instead, she developed gallstones that caused emergency gallbladder removal and life-threatening, acute pancreatitis. If we instituted universal genetic screening, it could find problems like this in infancy.

The ACMG recommends using their list of actionable genes when deciding whether to tell patients of incidental secondary findings that result from clinical sequencing for another disorder. The ACMG does not endorse using their list for general proactive population screening. They are concerned that some mutations in genes on their list will cause disease only in certain populations.[6] For example, a monogenic mutation that causes severe illness in one population might be less of a problem in another population that has a different array of background mutations in other genes. Population-based screening must take this into account. However, why this causes a distinction between reporting on disease mutations found incidentally vs.

INTERLUDE: PROS AND CONS OF GENETIC TESTING AND SCREENING

in population screening is unclear.[7] Possibly the reasoning is that for secondary findings the DNA sequence and counseling are already available, so it is more cost effective. ClinGen also falls short of recommending population-scale screening.[8] Their concern is that physicians are not sufficiently trained for such a large project, and that the follow-up required is not available. They do recommend reporting on incidental findings, which is on a much smaller scale.

Despite these problems, the ACMG is discussing a new list of genes and guidelines for universal screening. When—or if—such a list becomes available, it may encourage insurance companies to cover screening for the genes listed. This would save and improve lives.

Another new advance on the horizon is a move to have medical examiners promote genetic testing and counseling in cases of sudden deaths. The Research Triangle Institute International and Dr. Yingying Tang, director of the Molecular Genetics Laboratory in the Office of Chief Medical Examiner in New York City, told me they are starting such an effort. In addition, Dr. Tang has constructed a cardiac panel of 304 genes associated with sudden death for cardiac, epilepsy, aortopathy, thrombophilia, and sickle cell combined. She is using this panel to do postmortem genetic testing on old cases of sudden death. This is possible because some of the remains have been preserved without formalin, so she can isolate DNA. Using this approach, she has identified the mutations that caused sudden death in a number of cases. Genetic counselors in her office have been following up by informing and warning the families.

In the United States, it is illegal for health insurance companies to consider the results of genetic tests when issuing policies. Likewise, life insurance companies cannot alter existing coverage based on such information. However, they can consider genetic risk along with other medical data when issuing new life or long-term care policies.

CHAPTER THIRTEEN

It can be emotionally distressing to be aware of a higher risk of illness, even if there are treatment options available. Also, if revealed, this information may cause discrimination in social and employment settings.[9] It is imperative that we tackle these pressing concerns. The NIH National Human Genome Research Institute was established to promote the use of genetics to benefit human health while considering ethical, legal, and social aspect issues.[10]

FOURTEEN

Loss of Innocence

> "As for man, his days are like grass; like a flower of the field, so does he sprout." —Psalms 103:15-16.

It was November 16, 1980, and I was thirty-three years old. It was going to be a wonderful week. My parents were coming from Brooklyn to help me with the kids because Alan would be out of town taking a computer language class. Judith was two and Michael almost six. Also, that weekend Mom and Dad were going to look for a condo in nearby Evanston so their visits to us could be longer and more frequent.

After Dad and his business partner, Marvin, retired from their school jobs, Dad told me he would like to focus on growing their company. One large customer had come to dominate their business and had become demanding. Dad thought getting new clients would give them leverage with this bully. However, Marvin wasn't interested in opening a new chapter of work at their advanced age. After a few years, they contracted to sell the business to a broker couple who bought printing wholesale and marketed it. They wanted to use Dad's shop to expand and

CHAPTER FOURTEEN

do their own printing. Since they didn't have experience with presses, Dad and Marvin stayed on for five years as part of the agreement. After thirty years together, Dad and Marvin completed the sale and secured the jobs of their employees. Marvin and his wife went on to enjoy a long retirement on St. Thomas in the Virgin Islands.

A few days before my parents' visit, I got the news that the university granted me tenure. I was so excited! I had to call them. Alan wanted me to wait a few days and tell them in person, but I couldn't wait: I was bursting, so I called anyway. On the phone, I tried to explain why I was so elated that I didn't have a tenure hurdle ahead of me anymore.

After seven years as an assistant professor, you were either granted lifetime tenure and a promotion to associate professor or you had to leave. A committee decided whether you had published enough papers, brought in enough grant money, and had an engaging enough teaching style to commit to keeping you in the department indefinitely.

I saw many colleagues lose their positions because of this system. They often had trouble getting another faculty job and had to change careers. Since I thought I was ready for a promotion after four years, I had asked to have my case considered early. That way if they didn't grant me tenure I could still try again. But they granted it and promoted me to associate professor. Happy days. I was in!

Although my parents were delighted by my news, I'm not sure if they understood its full significance. However, the fact that the promotion brought me a sense of stability and reduced my worries made them happy. The academic accomplishment itself held little importance for them; instead, they prioritized my well-being, family life, and overall personal contentment.

The kids called my parents Nana and Papa. When they arrived, Mike and Judy ran to greet them. Papa said he and Nana were out late the night before at a wedding, which exhausted him. In response, the kids tucked him into the lower bunk in Michael's double-decker bed. We turned off the overhead lights to emphasize the blinking-colored lights set up in Mike's room while the kids danced to a Disney music tape. Judy idolized her brother and did whatever he told her to do. "Dansco Disko," as Michael called it, was their current favorite game. In between dances they joined Papa in bed and snuggled together with him. After Papa had rested for a while, I packed everyone into the car and drove to a nearby pancake house for a light dinner.

It was a family-friendly restaurant. As I adjusted Judy's bib, thoughts of a marvelous week, outstanding future visits, and an illustrious career flashed through my mind.

In my reverie, I heard Dad say, "You know, I am not feeling well." My head jerked up. I saw his gray face and hollow cheeks. How had I missed this before? "I think I need to go home," he continued.

What's wrong? Thoughts assailed me in a panicked muddle. Time to go home? We are still eating. What's going on? Frantically, I signaled to the waitress. "Please give me the check. My Dad isn't feeling well. We have to leave now."

Her eyes opened wide as she took a step back, looking at Dad. "What's wrong? Should I call an ambulance?"

Blood drained from my face and hands. "Dad, should she call an ambulance?"

Dad paused and shook his head. "No, no, just take me home," he said in a soft voice.

I looked back up at the waitress quickly. "No, just the check, please."

I drove with Mom in the front passenger seat. Dad was next to the kids in the back. As we neared the house, he spoke up loudly. "I'm really not feeling right."

I tried to focus on the road while processing this new information. "Should we continue going home, or should I drive to the hospital?" I asked him.

CHAPTER FOURTEEN

Dad answered quickly in a high squeaky voice. "Maybe you better go to the hospital."

Oh, my God. He wants to go to the hospital! How do I get there? It's not far. I don't want to get lost. We are only a block from the house. "Stay in the car everyone. I will run in and call for an ambulance when we get home."

The ambulance arrived within a couple of minutes of my call. This level of service amazed and pleased Dad. "They are here so fast. They would never come so quickly in New York!"

Dad was getting out of the car when a paramedic held him back. "Are you the patient? Don't get up. Let us put you on the stretcher."

Dad handed me his watch and glasses as they rolled him into the ambulance. "Hold these for me."

I ran up the stairs with the kids, dropped Judy in her crib, and called my neighbor to ask him to come over so I could go to the hospital. I also called my technician's father, a cardiologist, to ask for advice. He had none—just wished us luck.

By the time the neighbor arrived, the ambulance was gone with both my parents, sirens blaring. Nana called a few minutes later from Skokie Valley Hospital. "Come quickly. They say they are waiting for a doctor who can install a pacemaker. They keep asking me if someone is with me."

My hands shook so much that I had trouble opening my map to confirm the route. The hospital was only five minutes away. I parked in the Emergency Room lot, ran in, and found Mom. They at once ushered us into a private room where a doctor came to tell us, to tell us . . . to tell us . . . Dad was dead. He was only sixty-six, but by his own account, he had squeezed everything that was important to him into his short life.

His heart had stopped in the ambulance when they were still in front of our house. I had asked them to go to Evanston Hospital because it was a teaching hospital. But when they called, Evanston told them to go to the closer community hospital,

Skokie Valley, ASAP. En route, the paramedics tried to shock-start Dad's heart. The doctor in Skokie Valley also tried to restart his heart but to no avail.

"Did he have a heart condition?" the doctor asked with a clipped Indian accent.

Mom was trembling and breathing rapidly. "No, just angina."

"Angina, well then, it shouldn't be a surprise. It's a coroner's case though. They will do an autopsy to confirm the cause of death."

"No autopsy," Mom declared, wringing her hands, "desecration of a corpse is against the Jewish religion." It surprised me that Mom was so adamant about this since she was not a very observant Jew.

The doctor looked at us with his big dark eyes. "OK, I'll sign off on it if that's what you want. No autopsy."

A nurse came in and asked matter-of-factly if we needed to call anyone. I reached Alan and told him to come right home. Next, I called Diane. She knew immediately. "Is it one of them, or both of them?" she wailed.

How did she know? I couldn't restrain my arms and head from shaking, and I had trouble swallowing. "Just Daddy."

"Daddy is dead," she shrieked. Then I heard glass shattering. Bang, bang, bang, and she hung up.

The nurse asked if we wanted time alone with Dad. Mom preferred to remember Dad in life. But she wanted me to confirm that he was really dead. When I went in, I found a mannequin with a gaping mouth masquerading in Dad's clothes. Dad was gone. His body, a poor caricature of the man, mocked me.

Mom insisted we take a taxi home, fearing it wouldn't be safe for me to drive, so I moved my car out of the emergency parking lot. When I returned, Diane had called back to confirm that it was true. "Is Daddy really dead?" Jeffrey, thirteen, had heard Diane's lament and started throwing things at his large glass bottle collection. That was the glass-breaking noise I had

CHAPTER FOURTEEN

heard. He and Papa had been very close, and he was still howling in the background.

Mom and I lay in the same bed that night. It was a comfort to have her nearby. Nothing seemed real. Would the sun rise? Probably not. But the sun did rise. How was that possible? Dad was dead. The earth couldn't keep turning. Dead, like not coming back. Gone. Never again. The end. So this is it. This is what we get in the end. NOTHING. NADA. OVER. And then the sun goes ahead and rises like nothing at all has happened. What an insult.

I hadn't understood what death meant before that day. So naive. So innocent. Losing Grandma was no preparation at all. Useless.

Alan arrived and took charge. Jewish burials are supposed to occur within three days. To make that deadline, we flew to New York the next morning and arranged for the undertaker to receive Dad's body from the plane. That afternoon we picked out a casket and drafted an obituary. When I called the newspaper to place it, I made the mistake of letting out a single sob. "You better have your funeral director call us," the woman on the phone snapped, unwilling to put up with any display of emotion.

During the burial we noticed three more adjacent plots. We understood that one was for Mom but we asked her, "Who are the other two for?"

Mom looked at us soberly. "Whoever needs them."

Dad's death split my life in two. There was the "before," happy-go-lucky, naive me, "ignorant" of death and the pain it entails. And the wiser, much sadder, "after" me. For a decade, I could tell if something happened before or after Dad's death because I remembered if I felt like the "before" or the "after" me when it happened.

Our New York relatives, Diane's entire school, her friends, my parents' friends, and the Brooklyn neighbors all came to console us at my parents' apartment during the week after the funeral. This custom is called a shiva. My mother, sister, and I were the mourners "sitting shiva." The guests brought food, prepared meals, cleaned up, and tried to comfort us with their conversation and visits. Since I hadn't lived in New York for a long time, I hardly knew most of the people who came. Although it was nice to reconnect with long-lost relatives, I also needed my current friends and community. They were back in Chicago, although my most important support—Diane, Alan, the kids, and Mom—were with me.

A common theme at the shiva was amazement that my mother was burying my father. All the longtime friends and relatives remembered Cyrilla as being critically ill and slated to die young. She had cheated death. How did this happen to Norman, who was always the picture of health?

The first year was incredibly painful. My sister, Mom, and I spoke on the phone every day. For a long time, we agonized in the "if only" stage. "If only" they had taken him to Evanston Hospital instead of Skokie Valley . . . "If only" I had called an ambulance from the restaurant . . . "if only" . . . "if only" . . . "if only" . . . then he would be alive today. We seemed to believe that if we found the right "if only," we could reverse time and bring him back to life. Eventually, we came to accept that no "if only" mattered. Dad was dead and nothing could change that. From then on, I continued to talk directly with my sister, although our daily call eventually became weekly.

In time, when I woke up in the morning, I had a few seconds when I was free of grief. For those few seconds, I forgot. It just seemed like a nice new day with no worries. And then . . . all at once, I would remember. A tsunami of grief would knock me down as if I were learning of it for the first time. I realized that this torment, this unbelievable agony, was a normal part of life—the most you could hope for, an adult child losing a parent. How could I have not known this terrible secret, the meaning of death?

CHAPTER FOURTEEN

With the pain and despair I was battling, I couldn't imagine how children, like my father and Cyrus, could have survived the loss of a parent. How did Marion and David withstand the loss of their four-year-old son Eugene, or my great-grandmother Elke the death of six of her children?

According to Jewish custom, after sitting shiva for a week, mourners are to return to work. It was important to have that guideline, because at the end of the week, I didn't feel ready to return. Even though I went back to work, our tradition barred me from participating in entertainment during the first month of mourning. I didn't watch movies or television, play games, or go to parties. I realized later the value of avoiding these diversions. They could have caused me to temporarily forget my loss only to then face the excruciating wave of grief upon rediscovering it. During this month, I still wore the black ribbon the rabbi tore at the funeral as a symbol of my rending my clothes in anguish. I pinned that ribbon on each morning, wearing it as a badge to remind everyone of my grief. This comforted me. It let me feel it was OK to be despondent. I wasn't ready when, at the end of the month, I had to abandon the ribbon and move to the next stage of mourning. I no longer looked like a mourner without my ribbon. Was I supposed to feel normal? Was this the new normal?

Orthodox Jews expect sons to say the memorial prayer, Kaddish, three times each day for a deceased parent, for eleven months. This is supposed to help lift the parent's soul into heaven. Since my father didn't have sons, we hired the cantor of the synagogue we belonged to in Skokie to say Kaddish for us. This type of substitution was customary. The cantor carried Dad's name in his pocket when he said the memorial prayer.

Although we belonged to a congregation nearby, we were not active participants and weren't comfortable with the members or the rabbi. I realized I should belong to a congregation that could support me if another torment hit, or on a happier note, help us celebrate family simchas. Within a few years we found a new congregation, where we celebrated our son Michael's becoming a bar mitzvah. We became involved with this community and got to know the rabbi and his wife well. But when my father died, I didn't have the comfort of this rabbi or congregation yet.

The important thing about the Kaddish prayer is that you can't say it alone. Rather, mourners must recite it with a group of ten or more Jews, called a minyan. In our Conservative branch of Judaism, women count in a minyan and can say Kaddish for their parents—although few do it for an entire year. I said Kaddish several times a week at the synagogue's daily services for the first month. The people in this weekday minyan were all recently bereaved, so it was like a grief group therapy session.

Since I spoke with Dad only once a week when he was alive, I couldn't understand why now I missed him every second of every day. I tried to make believe he was in New York as usual. That way, I reasoned, I should just be sad when I couldn't reach him on the phone on Sunday. This didn't work. There was no way to trick grief. Time was the only remedy. It would be about ten years before I became whole again. Ten years until I stopped measuring everything in terms of whether it occurred before or after Dad died. During those ten years, I worked, played, loved, mothered, and celebrated, but beneath it all, I grieved.

I had never had much trouble with the commandment to not be jealous. I knew it was destructive. That changed after Dad died. I became exceedingly resentful of everyone whose father was alive, especially if their father was older than sixty-six. I was also particularly envious of people whose fathers survived heart trouble. If someone told me their father had triple-bypass surgery and recovered, instead of rejoicing with them, I wondered, why didn't my father have that surgery? Why isn't my father alive? I knew this was wrong. I was ashamed of this reaction. But there it was. I continued with this hurtful jealousy for many years. Finally, finally, finally, time healed me.

CHAPTER FOURTEEN

I never understood college essay questions such as, "If you could have dinner tonight with anyone living or dead, whom would you choose and why?" Answers like Einstein, Moses, and George Washington make no sense to me. Of course, I would choose to have dinner with my dad. I would choose him in a heartbeat.

Even though he would be over one hundred now, I still imagine "if only." If only he had lived. I imagine him here with me now, as loving and young as the day he died—younger than I am now. He is thrilled to see how my children, Judy and Michael, turned out. He is proud of their college and graduate school successes and their careers. He is happy to meet their spouses, Ben and Susan. He is awed by their children, Aaron, Amy, Adam, and Mark.

"Susie, I couldn't have wished for more for you. I am so content with your journey. Tell Aaron, Amy, Adam, and Mark to aim to be satisfied. If they can achieve that, they will be blessed. Also, remind them to do their best and leave the rest. And tell them I have full confidence in them."

FIFTEEN

The Aftermath

"It ain't over till it's over." —Yogi Berra

Mom's soulmate and partner for over forty years was gone in an instant. Totally unexpectedly. She had dealt with death in the past. She had been in the grips of its black hole before and escaped. She had seen others stripped of their partner, half of their former selves, somehow learn to be whole again. She was sixty-five. She was a survivor. She had more living to do.

When Diane and I thought we couldn't go on, when we saw our future clogged with these chasms, Mom consoled, "Part of what is so painful for you is that this is the first time you are facing death. I felt like that when my father died. I promise you, next time"—presumably her death—"won't be as hard."

Dad had taken care of many things Mom now needed to learn about. Although this was years before Alan became an investment advisor, he already had interest in and knowledge of finance. He worked with Mom on her budget, going over her savings accounts and investments. Dad had a black loose-leaf notebook with careful records, making this easy. Mom also wanted to learn about charity. She knew Dad was generous, and she wanted to continue to support worthy causes.

Dad was the one everyone counted on. He was always there with calm advice. He learned about rock 'n' roll to connect with his son-in-law Ronnie. He taught me how to drive on the highway. He played with the grandkids. Mom knew the family depended on Dad. When he died, she tried to fill his shoes as best she could. But seeing her out on the street, playing catch with our son, Michael, was a shock!

I came to realize that some of Dad's good deeds had come from Mom. I now remembered hearing her with encouragement and suggestions such as, "Michael looks bored, Norm. Why don't you go out and play ball with him?" She and Dad were more of a team than I had realized. Mom liked to stay in the background, but she enabled Dad to be who he was. She was content to have people award him the glory.

With Dad gone, Mom visited us in Skokie often. She loved lectures and classes for seniors at Northwestern University and the Jewish Community Center (JCC). She made two good friends there. One of their sons was a professor of Jewish studies at Northwestern and a regular speaker at the adult education lecture series they frequented. Now, instead of Mom taking cabs to class, her friends picked her up and dropped her off.

About a year after Dad died, Judy, aged two and new to daycare, ran off to her cubby. She sat there holding her blankie and sucking her middle finger much of the day as she studied her surroundings. I dropped her off at the JCC at 9 a.m. before my forty-five-minute drive to the University of Illinois at Chicago (UIC). Her eyes were open wide in amazement as she watched the other kids playing. Abraham, her special friend, marched back and forth in front of her, making funny faces that caused her to giggle.

As I left, Abraham's parents called after me. They were also faculty at UIC. "Can Barry get a ride to school with you today? My back just went out so I'm taking the car home now instead of driving to work with him."

I squinted. Although I was happy to have company on the long drive, this didn't sound good. "Sure. I hope your back is OK, though."

Carmel sighed and wrinkled her forehead. "I do this now and then. I'm usually fine after a few days' rest."

Barry and I settled into the front of my green, four-door Ford Maverick. We spoke of our kids and the daycare. Even though we were both at UIC, our fields, biology and economics, were very different. I told of the difficulties of getting grants, Barry of the many courses he taught while publishing books and presenting at conferences.

As I scooted into the express lanes, I asked, "Did I see you and Carmel last week at a Congregation B'nai Emunah meeting? Are you members there too?"

Barry sat up tall and faced me. "Yes, that was us. We're members. I didn't see you. We had to leave early that night to arrange something for my mother. She lives in New York City. She's blind."

I pursed my lips and furrowed my brow. "Is she familiar with the Lighthouse? My dad used to volunteer there."

Barry was taken by surprise. "Yes. She loves that organization. What's your father's name? I'll ask her if she knows him."

I converted a sob into a sigh. "My father's name was Norman Weiss. He died suddenly a little less than a year ago."

Barry's deep voice was even deeper when he responded. "Oh, Susan, I am so sorry for your loss. Tell me about him."

I smiled despite the enormous lump in my throat. "Thanks, Barry. I love to talk about him. It's very therapeutic. My father was only sixty-six when he died. He had recently retired and that's when he started volunteering at the Lighthouse. Dad was

CHAPTER FIFTEEN

concerned about people who went blind when they were too old to learn braille. He wanted to invent something that would enable them to read. He knew a lot about paper and ink because he owned and ran a print shop before retiring and talked about using inks with different textures. The first thing he wanted to do was get to know blind people who didn't know braille and their needs. He told me that people often mistakenly group everyone who is blind together in their minds. He said, 'The blind are just as diverse as the rest of us. Some are conservative, others liberal.' Dad realized the people in his class wanted to be in touch with the world, so instead of presenting a classic novel, he decided to read them the newspaper followed by an open discussion. This approach rewarded him with a number of friendships he valued."

"My mother would have treasured something like that," Barry said. "How is your mother adjusting?"

"She is visiting us this week and working hard at establishing a new life." As we approached Barry's building, University Hall, I asked, "What time should I pick you up for the return trip?"

"Thanks for the offer, but I'll catch the train home. I'm planning to leave early."

I generally left school at 5:45 p.m. to miss the worst of rush hour. Alan, who worked near the house, picked Judy up and prepared dinner. That night the trip home was a ninety-minute disaster. There was a pile-up of twenty cars. There weren't any cell phones in those days, so I couldn't call. When I arrived, my mother was in a panic. Alan said she stood by the window looking for me for an hour holding on to the draperies and whimpering.

As Alan herded everyone to the table for dinner he said, "Carmel phoned a little while ago. When Barry told her about your conversation, she remembered her mother-in-law talking about a special man who ran a current events class she attended at the Lighthouse." Alan paused as he lifted Judy into her highchair and put on her bib. "Apparently when Barry and Carmel moved to Chicago, Barry's mother gave them our contact information that she got from your dad. Carmel knows it was us because she just retrieved the slip of paper from her bulletin board. Barry had already called his mother to tell her about the coincidence. She was very sorry to hear that your father died. The Lighthouse didn't inform his class of his death. She sends her condolences."

Mom was listening with rapt attention. Her eyes were tearing but she was smiling.

Seven-year-old Judy's face lit up and her eyes widened. She turned to me and said, "Mommy, look at the size of that birthday cake! I hope Nana will like it." It was Saturday, March 16th, and my mother would turn seventy that Monday. She had been a widow for five years. Not wanting us to miss this chance to honor Mom, Diane had reserved the community room in her building and planned a big party.

Although Diane's daughter, Karen, was only thirteen, she looked like a model as she decorated the room with Diane's friends. "Judy, can you help us put the flowers on the tables?" Karen invited. I adjusted Judy's new dress and she ran off.

Diane's son, Jeffrey, a senior at Midwood High School and co-captain of the swim team, towered over me. "Aunt Sue, I'm taking Michael upstairs for a few minutes to show him Grandpa's track medals. He also wants to see my swimming trophies."

My son, Mike, was eleven and loved spending time with his big cousin. Michael was considering what sports he would like to play in high school. "That's fine. Just don't be late for the party," I said.

My husband, Alan, and Diane's husband, Ronnie, soon arrived with the catered food. We had open seating for about thirty guests, including my uncle Herb and his family, Mom's

CHAPTER FIFTEEN

cousins, and Diane's friends from the neighborhood. One of Diane's friends taped the highlights of the party with her video recorder. Another took pictures. There was a microphone on the dais. Both Alan and I gave speeches and toasts. We sang "Happy Birthday" to Mom and then again to her cousin Edith, who shared Mom's birthday. Judy, the youngest, got the first piece of cake. This delighted her. It was a splendid party that reminded Mom how much we loved her.

The next day Diane, Mom, and I had time to visit before my family had to leave for the airport. Sitting on Mom's couch, I wrested my attention away from the spellbinding view of the Brooklyn Bridge and faced my sister. "Congratulations on Jeffrey's being elected swim team captain."

Diane put down her coffee. "Thanks. He loves the sport. It was brilliant of his Midwood coach to make him co-captain with his friend. This way they both got to tout the position in their college applications."

Mom laughed. "I can't believe I'll have a grandson in college next year. Before long, he'll be a lawyer like his dad."

Leaning my chin on my clasped hands, I said, "Michael is interested in track. Jeffrey showed him Dad's medals and Mike is a fast runner. It could be a good sport for him."

Mom jerked her head up. With a quavering voice, she said, "Don't let him run short-distance track. That's what killed Dad."

My mouth fell open. I hadn't heard this before. Previously, Mom had expressed pride over Dad's athletic success. "Exercise is supposed to be good for your heart. What makes you say it was bad for Dad?"

Mom cast her eyes down. "Doctors told us Dad's heart was too large. They said the short-distance running he did in high school and college enlarged his heart muscle and that when he stopped running the extra muscle became flabby, which was dangerous. I think that's why he had the heart attack." She was weeping now, "Please don't let Michael run short-distance track."

In the end, Michael took up tennis. But years later we learned that running wasn't the cause of Dad's enlarged heart after all.

Although Mom didn't travel abroad after Dad died, she vacationed with her cousins in Chautauqua in upstate New York, where she enjoyed plays, concerts, and lectures. When she was seventy-one, and six years a widow, she met Samuel Harrow there. They married in 1986 in city hall, with no religious ceremony or family present. Sam got along well with my sister and me, but he put an end to Mom's talking with Diane three times a day. I think that helped Diane and her family.

In 1989, while out walking in her neighborhood in Brooklyn, Mom fell and bruised herself badly, but she didn't tell Diane or me about it. A few months later, the pain in her hip got worse and she had to go to the hospital. I was with her when the doctor told us she had a hairline hip fracture. He asked when she had fallen. As she answered, she wore her guilty grin with a telltale flick of her tongue toward her nose. She knew she should have gone to the hospital right after the fall. She had tried to pull a fast one, hoping to recuperate at home. I don't remember if they asked why she fell. In retrospect, a mini-stroke likely caused the accident. I wish her doctors had realized that and had put her on blood thinners.

Instead, Mom had a major stroke a few months later when she and Sam were in Sun City West visiting Mom's brother, Herb, and his wife. Sam called 911 when he found her unable to move or talk. Our aunt alerted Diane and me to come right away. As soon as Mom saw me, she became agitated in her hospital bed and tried to tell me about the terrible thing that just happened to her! But, alas, she had no words.

We hoped she would regain speech and movement with physical therapy. Diane and I took turns traveling to Arizona

CHAPTER FIFTEEN

to visit and supervise her care. Sam stayed with her much of the time. She couldn't swallow, so they inserted a feeding tube through her nose. Diane and I wanted them to continue physical therapy. We weren't ready to give up. It seemed to us she responded to things we said. However, the doctors insisted these responses were just reflexes. Before long, they refused to place orders for therapy and insisted we move her out of the hospital into long-term care. We managed to get her a bed in a nursing home in Brooklyn where Sam and Diane could visit often. However, Mom got an infection almost at once, and they transferred her to a hospital.

About then I asked our rabbi for advice. This was just before Passover. "Should I let Judith, age twelve, and Michael, age sixteen, visit my mother in the hospital? Should we sign a do-not-resuscitate order?"

The rabbi was direct and firm in his advice. "Yes, certainly your children should visit their grandma in the hospital. They need to say goodbye and learn about death. Also, absolutely sign a do-not-resuscitate order. You need to do this to let your mother avoid as much pain and suffering as possible."

A few days later, we came to Brooklyn to celebrate Passover with Diane's family and to visit Mom. By then my mother was totally non-responsive. Her husband, Sam, had followed Diane's and my wishes and signed a do-not-resuscitate order. Mom died the next week. She was seventy-four.

The burial and shiva took place in New York. This time, Alan asked our rabbi, as well as several of my friends from our Chicago congregation and my college friends Razel and Ellen, to call me during the shiva. This comforted me.

Unlike the shock of my father's death, we were more prepared to lose Mom, since her demise followed months of incapacitating illness. Also, Mom was right. As much as Diane and I missed her, this second loss wasn't as painful as when we had lost Dad ten years earlier, because death was not a stranger this time. Although we were now "orphans," we had built a close sibling relationship, and this helped us keep our parents alive in our hearts. Mom was at peace now lying next to Dad.

Later, when we were searching for the source of the family genetic disease, we thought the mutation could have caused Mom's stroke. Another possibility was that the scarlet fever that Mom had as a child attacked several organs, causing both her stroke and her earlier nephrosis.

SIXTEEN

The Mitzvah

"A mitzvah is a commandment and obligation, not a good deed."
—Rabbi Donniel Hartman

In the 1980s, when I was an associate professor at the University of Illinois at Chicago, it became difficult to attract grant funds. While I enjoyed thinking about ideas as I wrote the grant applications, I hated the pressure and anxiety associated with getting good reviews. One comfort was that I knew I could always quit and become a housewife like all my neighbors. Subtracting taxes and childcare costs, my income was negligible. Science was just a lovely hobby.

Alan's income supported the family. He worked at Teletype, a subsidiary of AT&T in Skokie, on printing technology. He liked his job and headed an excellent team that invented and patented an offset copying technique. Teletype paid him well, and the whole family benefited from the company's strong health insurance plan. However, the old AT&T monopoly had a surplus of managers from unrelated parts of their business empire and kept transferring those managers into jobs in research and development

CHAPTER SIXTEEN

where they didn't have expertise. In trying not to demote its surplus managers, AT&T created a barrier against the promotion of suitable would-be managers like Alan. Therefore, they passed over him for promotion a few times. With his recent successes, we were certain that this time they would give him a promotion.

I was unprepared when Alan entered the house and announced, "I quit my job today."

Putting down the lettuce I was washing, I turned to face him. "What do you mean you quit your job?"

Alan walked toward me, lifting his chin and crossing his arms in front of him. "They hired a transferred manager again. I told you, and I told them, I would quit if they didn't promote me this time. So I quit."

Is this really happening? Could he have taken this step without consulting me? Resting my forehead on my palms, I said, "I didn't think you meant it. You haven't even begun to look for a new job. I was sure you would line up another job before quitting this one!"

Alan pounded the kitchen table with his fist. "I don't want another job. I have had it working for big companies. I want to work for myself."

Imagining the bleak new reality, I wrung my hands. "What will we live on? What will we use to pay the mortgage? What will happen to the house? What about our health insurance? Oy! Oy! Oy!"

Alan put his arms around me. "Relax! Relax! Nothing will change. Your salary will cover our bills."

I shrugged. "But my job doesn't pay anything."

Alan smiled and steepled his fingers. "Yes, it does. I've thought it all through. Your salary just looks like nothing because with my income we have a huge tax bill. Once my income disappears, taxes will go down, and you will earn enough to pay the bills."

I clenched my fists and bit my lip. "Oy! Oy! Oy!" I worried about the ramifications of the lost income but was more concerned that Alan would get depressed and lazy at home with nothing to do.

I needn't have worried about Alan's resolve. He got up early each morning, put on a suit and tie, and took Judy across the street to wait for the school bus. For the rest of the day, he sat at his desk outlining ideas for a home-based business. He was writing grant proposals to himself. This was in 1984 soon after both our fathers had died.

Alan was right. My salary covered our bills. Also, we were able to switch the family to health insurance provided by the university. Nonetheless, I remained fearful and economized as much as I could. I stopped buying lunch in the school cafeteria and instead brought a sandwich from home in a brown paper bag.

Gone were the days when my work was a hobby. Gone was the feeling of power that I could quit at any time. Instead, I had the great joy of knowing my work was supporting my family. Instead, I knew I was doing something of value for the family. Instead, I wasn't guilty of playing a selfish game by working. Instead, I gave my husband a chance to find a new career. Instead, by being a scientist and professor, I was also being a good wife and mother!

It surprised me to find that my job was more meaningful and valuable when it became a requirement. I liken this to the value of the 613 mitzvot in the Jewish tradition. People think a mitzvah is a good deed you do voluntarily. I learned from Rabbi Donniel Hartman that doing a mitzvah is not optional at all. God commands us to do mitzvot. For example, one commandment is that we care for our parents. We must do this whether or not we want to, whether or not we love our parents, and whether or not they were good parents. It is an obligation, a commandment we are to fulfill.

AT&T shut Teletype down two years after Alan left the company. His elective departure saved him from involuntary job loss. Alan's first business venture was to write and sell an investment newsletter. When that proved disappointing, he began an

CHAPTER SIXTEEN

investment advisory company and named it "American Superior Company" after my late father's business, "Superior Offset Lithography." Alan ran this one-person business from our home for about forty years and earned more money than I. But I still felt that my job was important for the well-being of the family. My work remained a mitzvah.

Every scientist is different. Some are more focused than others, but this may not increase their scientific success. I tried to explain this to my daughter when she was a Stanford undergraduate majoring in engineering. She became despondent about her future because she didn't read engineering books for fun, preferring instead to go to a movie on a Saturday night. She worried that all first-rate engineers would prefer to study engineering day and night over other entertaining activities.

In my experience, it is just the opposite. Getting away from one's subject by going hiking, skiing, dancing, or to the movies frees up the mind and allows the scientist to be more creative. For example, Nobel Laureate Kari Mullis discovered the polymerase chain reaction (PCR), an ingenious way to amplify DNA, while he was driving to his regular weekend getaway. He also spent two years as a fiction writer after he got his PhD before returning to science.

Another Nobel prize laureate, Jack W. Szostak, was not always sure he would stay in science. I knew him when we were both postdocs in Fred Sherman's lab, where Jack completed trailblazing work. During that period, he talked about leaving science and its stresses to follow his passion to become a carpenter. Jack stuck it out in science and, in addition to winning a Nobel Prize, one of his graduate students, Jennifer Doudna, went on to discover CRISPR, a precise gene-editing technology.

CRISPR has many revolutionary uses, including the potential to transform the lives of patients who have harmful mutations.

The other distinguished scientists I know all have a life beyond science, including three of my close colleagues who were members of the National Academy of Sciences. My mentor, Fred Sherman, pursued ballet for decades and was serious about lessons and practice. My collaborator Susan Lindquist treasured ballroom dancing, literature, and theater. My colleague Reed Wickner is an avid sailor and enjoys racing. All three reveled in their children.

I love science and find it extremely rewarding to discover something new even now in the twilight of my career. I also love painting, reading novels, writing, watching movies, having dinner with friends, and spending time with family. And I am the most creative with research ideas after I get away from science for a while.

SEVENTEEN

Interlude: The Central Dogma of Molecular Biology and Me

"A righteous man falls down seven times and gets up."
—King Solomon, Proverbs, 24:16

One problem with getting my grants funded in the 1980s was that the protein synthesis subject I worked on was no longer fashionable. Instead, the scientific community wanted to know how specific DNA sequences in the front of genes, called promoters, turned genes on or off. This topic was exciting because they knew enough about promoters to study them on a molecular and DNA sequence level.

My lab had isolated mutations that hinder the cell's protein synthesis machinery from recognizing termination codons in DNA. Most codons tell the cell which of the twenty building blocks in proteins, called amino acids, to add next to make a protein. Termination codons tell the cell not to add any more amino acids but to instead release the completed protein from the protein synthesis machinery.

Some of the mutations we isolated altered the ribosome, a large machine in the cell that makes proteins. We wanted to know how the ribosome recognized the termination codon signals for the ends of proteins. We knew so little that our methods were crude. If I were to attract grant money, I needed to modify my approach. Thus, I proposed using my expertise in protein synthesis to challenge a prevailing theory about ribosomes.

Science is the study of hypotheses. If we can't disprove a hypothesis after many attempts, we come to believe it is true. In 1957, Nobel Laureate Francis Crick coined a fundamental hypothesis in molecular biology that he called the central dogma. He described the cell as a factory designed to make proteins. Some proteins have a mechanical function such as helping the cell keep its shape. Other proteins, called enzymes, make it easier for specific chemical reactions to occur. Still other proteins, such as those in the cell membrane, help the cell interact with the outside world. In humans there are also proteins with special functions in our organs. For example, proteins make up the muscle in our hearts that allows it to pump blood.

The central dogma proposed:

1. Deoxyribonucleic acid (DNA) encodes instructions for everything using only four "letters."
2. An enzyme transcribes (copies) the DNA code into messenger ribonucleic acid (mRNA), which uses four RNA letters that are like the DNA letters.
3. Ribosomes translate the mRNA codons into the language of proteins. Codons are three ordered letters. Each codon codes for one of the twenty amino acids or for the termination of protein synthesis. The order of amino acids in a protein determines how the protein folds to accomplish its function.

According to the central dogma, DNA codes for mRNA, which codes for protein. Proteins fold into unique shapes that allow them to do their specific job for the cell and organism.

INTERLUDE: THE CENTRAL DOGMA OF MOLECULAR BIOLOGY AND ME

Cracks requiring modification of the central dogma appeared in the 1970s. Information did not always go from DNA to mRNA to protein but could instead go from mRNA to DNA. Also some RNA folded into specific shapes and had an enzymatic function in the cell like a protein.

Ribosomes are made of many proteins and several specific RNA molecules (called ribosomal RNA). Because of the central dogma, researchers believed that the ribosomal RNA was just a platform orienting the ribosomal proteins so they could carry out the ribosome's function of translating mRNA into protein.

I and others hypothesized that the ribosomal RNA might not just be a scaffold for protein but might have functional activity itself, such as recognizing termination codons. We proved this by isolating mutations in ribosomal RNA that affected termination codon recognition. Grant reviewers judged this ribosomal RNA work groundbreaking. I was back in the game.

With grant money in hand, I recruited Dr. Yury Chernoff, a postdoc from Russia, to work on this project. Except for his Russian accent, he reminded me of Abraham Lincoln, with his tall lanky frame and beard. Yury had a long history of using genetics to study how the cell finds the ends of genes. He brought an unfinished project with him that involved a mysterious element called [*PSI+*].

In 1965, Oxford University professor Dr. Brian Cox showed that the [*PSI+*] element made it hard for the ribosome to find the ends of genes. He also proved that although yeast daughter cells inherited [*PSI+*] when they budded off from mother [*PSI+*] cells, DNA didn't code for [*PSI+*]. This interested just a few researchers worldwide, members of the "club" that worked on the cell machinery that finds the ends of genes.

Yury worked on [*PSI+*] in an earlier postdoc internship with Dr. Bun Ono in Japan. Bun and I trained together in Fred Sherman's lab in Rochester, which is how we both became interested in [*PSI+*]. In Bun Ono's lab, Yury found genes that caused the loss of [*PSI*+] if they were overexpressed. One such gene encoded a protein called HSP104 (heat shock protein 104). Yury continued this project in my laboratory. Dr. Susan Lindquist, a distinguished colleague at the nearby University of Chicago, was an expert on HSP104, so we enlisted her help. Together, we showed that cells lost [*PSI*+], becoming [*psi*-], when HSP104 was either overexpressed or was missing. All this time, none of us knew what [*PSI*+] was or that this project would forever alter the direction of each of our research careers.

My first inkling that we were on to something pivotal came when Dr. Reed Wickner, a researcher from the National Institutes of Health, called to tell me he had figured out that [*PSI*+] was a prion. This was especially exciting because prions were another challenge to the central dogma. Dr. Stanley Prusiner, a UCSF professor, proposed that an infectious form of a protein, which he named a prion, caused certain neurodegenerative diseases, including mad cow disease. The term *virion*, which means an infectious virus particle, inspired the name prion, meaning infectious protein. Until then, scientists thought infectious diseases were all caused by organisms with DNA or RNA genetic material, such as bacteria, viruses, or fungi. Dr. Prusiner proposed that pure protein could also be infectious and could even self-replicate. This was heresy!

The Nobel Committee awarded two Prizes for work on prion diseases: Daniel Gajdusek won it in 1976 and Stanley Prusiner in 1997. However, Prusiner's concept of an infectious heritable protein, the so-called "protein-only prion hypothesis" remained controversial in 1994 when Wickner called me to share his evidence that [*PSI*+] was a prion in yeast.

He revealed that the [*PSI*+] factor was an aggregated form of the protein named SUP35, which recognizes termination codons at the ends of genes. Wickner proposed that when SUP35 proteins aggregated to form [*PSI*+] they lost their ability to find the ends of genes. He hypothesized that the [*PSI*+] SUP35 aggregate continued to attract (or seed) other SUP35 protein molecules to join the aggregate. Once aggregated, these molecules themselves became seeds, attracting (or infecting) new SUP35 molecules to join the aggregate. Therefore, [PSI+] was like the controversial mad cow disease prion described in mammals: It was an infectious and heritable protein. This was particularly relevant to me because I was one of the few experts on SUP35.

HSP104 is a chaperone protein that helps other proteins stay out of trouble. Our work showing that HSP104 affects the ability of [PSI+] to propagate supported the hypothesis that [PSI+] is an infectious aggregated protein (i.e., a prion). This was important evidence for the general protein-only prion hypothesis. We now know that the [PSI+] prion forms fibers that grow from the fiber ends. HSP104 severs the fibers, thereby creating more growing ends.

Following the publication of Wickner's 1994 seminal paper in the journal *Science*, our work on the effect of the HSP104 chaperone protein on [PSI+] went from being of specialized interest to having a wide-ranging influence. We didn't set out to understand prions; instead, our discovery came from serendipity and simple curiosity about unknown results. This is often how scientists make important discoveries.

In the twenty-five years following the publication of our paper, many labs have looked at the effects of other chaperone proteins on different prions. They have also examined the effects of chaperones on prion-like aggregates associated with diseases such as Alzheimer's, amyotrophic lateral sclerosis (ALS), inclusion body myositis (IBM), and protein aggregate myopathy. Evidence suggests that chaperones may become a useful therapy for these diseases. To date, over one thousand published papers have referenced the work that we published on HSP104 in 1995 in the journal *Science*. It is the most important contribution of my career.

This paper also influenced my late collaborator, Susan Lindquist, to focus her work on prions using the yeast model.

CHAPTER SEVENTEEN

We had joint laboratory meetings, which furthered research on yeast prions in both groups until she left Chicago to become the director of the Whitehead Institute at MIT. This position made her one of the first women in the nation to lead a major independent research institute.

My lab has continued to study prions and prion-like proteins involved in diseases such as Alzheimer's, Huntington's, and ALS. In 2019, I picked up a side project studying the mutation that caused sudden death in my family.

EIGHTEEN

Empty Nests and Full Hearts

"Men plan and God laughs." —Yiddish proverb

When my youngest, Judy, left for college in 1996, I turned to my sister for solace and advice during our weekly telephone calls.

"How was the weekend at Stanford?" she asked.

Despite my loneliness, I smiled as I settled into my chair to tell her the story. Alan had driven Mike to start graduate school at MIT while Judy and I flew to her freshman orientation. "I rented a car so I could shop for Judy's linens and other dorm room necessities. Driving from the San Francisco airport I got in the wrong lane and accidentally crossed the bay on the San Mateo Bridge. I couldn't believe the bridge went on for seven miles!"

Diane laughed. As a seasoned driver, she rarely took wrong turns. "How does my brilliant sister get into such messes? How did Judy handle your detour?"

I sighed. "She stayed calm. Judy's accustomed to my freaking out when I get lost because I had meltdowns driving her to birthday parties when she was a little girl. She used to say, 'Mommy, it's OK if I'm late. Pull over and take out your map.' After we crossed the bay, Judy calmly guided me to take a different bridge, the Dumbarton, to get back. I waited in a long line to enter the bridge and finally held out my money to the toll collector. Instead of taking it, she waved me on, saying the car in front of me paid for us. I asked her why someone I didn't know would do that. She explained people do it to be nice. Can you believe a stranger paid for me and the toll collector didn't keep the money herself?"

Diane chuckled. "That would never happen in New York, California is nuts."

I arched my neck as I stretched. "The orientation week went well once we arrived." I switched the phone receiver to the other hand. "It shocked me to see men living in her dormitory even though I read about it before we came."

I heard Diane sip her morning coffee. "This is new. Karen's dorm was only for girls, but boys were always there anyway."

I ran my fingers through my short dyed-brown hair. "I wasn't the only parent upset about this. One father, a tall Black man, expressed my feelings perfectly during a question-and-answer session. Straightening his tie, he stood in the auditorium as he asked the administrators, 'Did I hear this right? Are you telling me that there are co-ed bathrooms in the dormitories? You're talking about my daughter here. I need to know. Is she going to be sharing a bathroom with men?' I wanted to stand up and clap for him. Both our daughters gained entrance to an elite university, but now we learned we had fallen down a rabbit hole and the school they entered was in Wonderland."

"Did you meet Judy's roommate? I hope they get along. I had a lot of trouble with my freshman roommate. It was one reason I transferred schools."

I tilted my head and smiled. "Judy and her roommate have a lot in common. The school paired out-of-towners with locals so I think Judy's roommate's parents will help look after her. I'm jealous of them. You're lucky Karen came back to Brooklyn after her graduation. How did you manage when Karen was away?"

To answer my question, Diane switched to her big sister coaching voice. "I started going line dancing several nights a week." Diane exchanged her pointe shoes and ballet from years ago for western boots and line dance in her early fifties when the legendary Manhattan dance studio Denim and Diamonds opened in 1993. Ronnie wasn't interested in dancing; he remained happily at home, listening to rock 'n' roll, reading the paper, and running in the park. One thing they did enjoy together was travel. They went on tours all over the world during every school break. Diane continued, "I love line dancing; I've got my bridge game and book club too. You can meet people playing cards. You could learn bridge or mah-jongg. You used to have a lot of friends, but now I'm the one with all the girlfriends."

I tightened my lips and shook my head. "You played bridge your whole life. I don't think I would like it. And I'm not a dancer like you."

"Fine, I didn't mean you should do the things *I* like. You should find things *you* like. What about drawing, painting, or ceramics? You used to love that."

I sat up straight in my chair. "Maybe I *should* take a drawing class. That's a great idea."

"You devastated Mom when you left for college. I remember her telling me in tears, 'It's not fair. Susan will only let us call once a week! When you lived at school, we spoke at least every day.' Mom's statement shocked me because it was the first time I understood she loved you too! Before thinking, I blurted out, 'You mean, you miss Susan that much?' She always made me feel like I was the center of her life with you of little importance. For years afterward I was dumbstruck that I had been so blind."

I giggled. "I certainly knew Mom loved me."

For the rest of the conversation, we tried to figure out why Diane thought Mom loved her more than me even though I didn't feel any favoritism. Our parents worked hard to be even-handed. When they paid for Diane's ballet lessons, they offered

CHAPTER EIGHTEEN

me ballet too. When I had little interest, they didn't force me to continue but let me know they earmarked the unused money for activities I might like better, such as art classes.

Their goal wasn't to make us equal. If one daughter needed something they could afford to buy, they helped her while saving an equal sum for the other daughter in case the tables turned. No matter what unequal curves life threw, I knew our parents loved us both and would respond equitably. One thing they couldn't control and didn't know was that at the time of our conception they gave one of us a terrible genetic legacy, while sparing the other.

Three years later, the ramifications of this legacy were about to make their debut. I curled up in the comfortable green upholstered armchair in the corner of our Skokie bedroom to prepare for my Sunday morning call with Diane. By then, Mom had been gone for nine years and Dad for nineteen so Diane and I relied on each other. I loved sitting there because the bright autumn leaves on the tree outside the window were at their peak and not yet falling. As I dialed, my eye caught subdued Halloween decorations on the houses across the street. "Hi. Happy pre-birthday!" Diane would be fifty-seven that Thursday. "How are you celebrating this year?" I expected that she would go out to dinner with her husband.

"Ronnie was planning to take me to a fancy restaurant, but I'm still not feeling well. I couldn't go line dancing again this week either. Ronnie is leaving a bowl of candy outside the door so we don't have to answer the bell for trick-or-treaters tonight. I usually dress up and hand out candy, but I don't have the energy."

I twisted in my chair and stared at the floor. "You are still sick?"

"Yes. There's been a substitute teacher in my class for weeks now. My girlfriend Bonnie says the kids don't like her. I miss my class. I hate catching colds from them. I've had colds, then bronchitis. This week the doctor diagnosed me with walking pneumonia because an X-ray showed fluid in my lungs and an enlarged heart. Bonnie found some stores with the best brand-name clothes at a big discount and wanted to take me shopping, but I was too sick."

It was always hard for me to get a word in edgewise when talking with Diane. "Did your doctor say the bronchitis turned into walking pneumonia or is it a separate infection?"

"We didn't talk about that. She didn't prescribe any medicine for bronchitis. Now I'm on antibiotics. They are making me nauseous. I loll around in pajamas all day. Ronnie tells me to lie down, but that makes it hard to breathe. I've never been sick this long." Without pausing and despite her labored breathing, Diane abruptly switched topics and started talking about Karen and Jeffrey.

After graduating from Tulane, Karen became an elementary school teacher like her mom. She earned a master's degree in early childhood education at night at Bank Street College. For a while she lived with an Iraqi Jew who worked in his family's diamond business. She felt sure they would marry, but that wasn't to be. Her being Jewish didn't satisfy his parents. They also wanted their son's wife to be Iraqi. Another problem was his jealousy over the attention Karen received because she was an attractive woman. She had an eye-catching figure, dirty blond hair, flawless skin, a straight nose, blue eyes, and a glorious smile. She also had a deep voice and an infectious laugh. Sometimes she had a nervous cough, but even that seemed attractive. After their breakup, Karen traveled in France for a while focusing on becoming fluent in French. She was a free spirit.

"Karen has taken up acting. She loves it and is pretty good. I'm lonely for Jeffrey and Laurie." Jeffrey's wife, Laurie, had recently completed medical school. And now, without skipping a beat, Diane switched the conversation back. "I can't even walk

CHAPTER EIGHTEEN

across the room without getting out of breath. We order takeout every night. Ronnie likes the food but hates to spend money. I'm not hungry."

I took a deep breath and gathered my courage to interrupt Diane's monologue. "What does Laurie say about your illnesses?"

"I told Jeffrey I was sick but didn't give details or ask for advice. I wish Laurie's internship weren't so far away. She'll make a wonderful doctor."

"Can you sit up in bed and watch television?"

"I try. My friends came over on Thursday to play bridge, but I couldn't concentrate. I wasn't able to go to my book club either. I wonder why Ronnie didn't catch any of my colds, bronchitis, or pneumonia. He wakes up at 4 a.m. and runs in the park at 6 a.m. every day. OK. I'm going to stop talking now. Jeffrey says I monopolize conversations and should learn to listen. Tell me something about you."

I lifted my head and smiled, thinking about my kids and putting concerns over Diane's health aside for the moment. "Judy and her boyfriend are still working on the solar car. Alan and I just made plans to drive to Indianapolis and watch them take off for Colorado Springs in Sunrayce."

I could still hear Diane panting. "And what about Mike?"

I sighed. "He's on track to finish his thesis soon. He works too hard."

Diane was gasping now. "Just like . . . you. I'm . . . glad you took . . . my advice and . . . are painting again. You . . . should also join . . . a women's . . . group."

I could no longer ignore her rapid, shallow breaths. "All this talking is tiring you out. We should get off now. I'll call tomorrow to check on you. Try to remember Mom's adage, 'This too shall pass.'"

Something was very wrong with Diane. Ronnie and Karen seemed to be in denial. I decided to alert Jeffrey in Baltimore. I dialed his number. "Hi, Jeffrey. I'm worried about your mother. She has had respiratory distress for over two months. First there were colds, then bronchitis and now walking pneumonia. She gets out of breath just going across the room. Also, I'm not confident that her pliable antibiotic-dispensing doctor is giving Diane the best medical care."

"My mom's really sick? She mentioned having pneumonia, but I didn't know it was so bad. I'll drive down next weekend."

Diane called me a week later. "Laurie took one look at me and said, 'You know, Mom, I think it might be your heart. You should see a cardiologist.' She thinks I have heart failure and is getting me an appointment with a cardiologist early next week."

This diagnosis frightened and perplexed me. "Oh my God, heart failure! Isn't that when your heart stops? How can you be alive if your heart stopped?"

"Laurie told me the name heart failure is a misnomer. The heart hasn't failed yet. It's having a hard time pumping and without help will stop."

Diane was weeping when I next heard from her. "They gave me an echocardiogram. Laurie was right. I am in heart failure. My left ventricle is too large. I have what they call dilated cardiomyopathy—or DCM—which causes heart failure. My heart only pumps out 19 percent of the blood in it with each contraction. This is life threatening because it should eject 50 to 70 percent of the blood." Her sobs made it hard to decipher her words. "I think I'm going to die soon. Heart transplants are one of the major treatments, but I'm not sick enough yet to be a candidate."

My jaw dropped and there was a lump in my throat. "Why, all of a sudden, is your heart having trouble? Where did this come from? You eat right and exercise all the time. You're only fifty-six. I don't understand."

Diane responded in a loud, angry voice while sobbing. "They don't understand either. That's why they call it idiopathic. They don't think smoking is the culprit since I haven't smoked for thirty years. I never took drugs and have no diseases that cause it. And I don't drink and I'm not overweight. Since no one else in the family ever had heart failure or DCM, they say I got it from a virus. They assume the virus that caused one of my colds settled in my heart and claim this can happen even from a minor illness." After she got control of her sobs, she added, "They don't think the heart damage is reversible. What if I die before I ever have a grandchild and before I see Karen married? I'm not ready."

I didn't know how to comfort her. What could I say? This was unbelievable. "Oh my God. I am so sorry. Is there a treatment suggested while you wait to become a candidate for a heart transplant?"

"He gave me a lot of pills to take several times a day including ACE [angiotensin converting enzyme] inhibitors and beta blockers. The ACE inhibitors relax my blood vessels, letting the blood flow more easily, even with my lousy heart. Beta blockers slow the rate of beating. I designed a chart with alarms and a system to ensure that I don't miss any pills or take a double dose, which could be fatal. I'm so scared. The pills won't make my heart pump any better, but they relieve my symptoms. He told me to sleep with a lot of pillows under my head and torso. That is helpful. I would have died without medication, but even with medication my heart is likely to keep getting worse until I need a heart transplant." She paused and then said, "I am trying to hold on to Mom's warning that many people become extremely sick in middle age as she did but that most recover. It is as if she saw into the future and was trying to arm me with courage."

After Diane's recovery from heart failure, her dilated cardiomyopathy kept her ejection fraction low (19 percent). The ejection fraction is the fraction of blood the heart pumps out of the heart with each heartbeat. Diane became reconciled to having only a few years to live and took disability leave from her job. Jeffrey and Laurie moved back to Brooklyn into Diane's co-op building and had a baby earlier than planned so Diane could experience being a grandma in her remaining years.

Karen's boyfriend, Don, whom she met in the New York City actor's workshop, moved in with her. He was a tall, thin man with deep brown hair, thick eyebrows, small smiling eyes, a mustache, and a goatee. Although there was no Judaism in Don's upbringing, he was technically Jewish because his maternal grandmother was Jewish. His mother, with 50 percent Ashkenazi genes, died young of DCM, coincidentally the same illness Diane battled. Don became estranged from his Puerto Rican father following a series of arguments and supported himself as a waiter. Although younger than Karen and still in college, Don had all the self-confidence Karen lacked.

Jeffrey and Laurie's baby arrived shortly after 9/11. They named him Noam after my father. Mom used to call my father Norm, so the baby's name was similar. His bris, a Jewish circumcision ceremony that occurs when the baby is eight days old, was so soon after 9/11 that most people still avoided flying. Diane urged us to come, so we gathered our courage. Since the event was on a weekday, we caught an early morning flight with a return later that night. We didn't bother with luggage, taking only a greeting card and a check for a gift.

During our taxi ride to the synagogue where the bris was to take place, we saw police everywhere. An officer guarded the door as we entered the building. We were early, so I had time to help my sister decorate tables in the large social hall.

Jeffrey and Laurie's colleagues came directly from work. Diane and Ronnie's friends and neighbors joined us too. The main event, the mohel's circumcision of the baby, took place far away from the guests. Someone whisked Noam away afterward. I don't recall if he fell asleep in his carriage at the other end of

CHAPTER EIGHTEEN

the room or if he went home with a babysitter. In either event, we didn't see him during the festive buffet dinner.

At the party, Jeffrey and Laurie were busy as hosts. I had an enjoyable conversation with my niece. As we stood together at the dessert table, Karen told me, "Did you know I'm leaving for France next week? I'm so excited. I'm going to teach English in a French elementary school, and I will even get paid. I'll be gone for the entire school year."

I chewed on some dark chocolate baby carriages as I responded. "Congratulations. How does Don feel about your going?" By then, Don and Karen had been living together for about two years.

"Don is fine."

"How is your French?"

Karen gave a shallow nervous cough before she started speaking in French as she twirled around and gestured theatrically with her arms. "I can get along in French fine, Aunt Sue, but I hope to improve it a lot during the year."

Now I moved on to the chocolate-covered mints. "Do you have an email address? I'd like to write to you directly instead of only hearing about you from your mother." As she wrote her address, I resolved to get closer to her and Jeffrey.

After all the chocolate I ate, my stomach was queasy on the return flight. I tried to settle it by asking the stewardess for ginger ale. That night, I woke up with a start at 2 a.m. I jumped out of bed terrified because I couldn't breathe. Unable to make any noise through my throat, I turned on a lamp and put my hand around my neck to signal I couldn't breathe. Alan woke and pounded on my back. That didn't help. Next, he tried the Heimlich maneuver. Also, useless. It seemed to go on forever, but I knew it would end soon. I was going to die.

Finally, a miracle. A little air got through. Desperately, I achieved a very shallow gasp. Then more air and a larger gasp. I might live after all. I had experienced a few similar but less severe episodes like this before. These lasted less than a minute and occurred while I ate chocolate mints. I had assumed I was allergic to something in the candy and that my throat closed up upon eating them.

The next morning, I called my allergist and insisted I had to see him right away. During my appointment, he listened to my tale and asked, "Did you eat anything out of the ordinary yesterday?"

How did he know? "Yes, I ate a lot of chocolate, some mints, and soda that I rarely have."

He shook his head and smiled knowingly. "You had a laryngospasm. Eating chocolate, mints, or carbonated beverages relaxes the sphincter between the esophagus and stomach, allowing the larynx to sense stomach acid. Your larynx is unusually sensitive and went into spasm to protect your lungs. Lying down makes this happen more, so sleep with pillows. When it happens, it will only last a few minutes, although it will seem longer. Also remember, it won't kill you unless you die of fright."

I continue to suffer from laryngospasms. Trying to breathe out, rather than in, helps. It also helps to avoid my triggers like vinegar. I remind myself that the doctor said it will not kill me, but I still get terrified.

When I got home from work that evening, an email from a friend in my Jewish women's awareness group was waiting for me. Email was new then. We were still learning of its power. This note warned of disenfranchised Maghrebi (North African Arab) youth assailing Jews in France. The information seemed relevant to my niece because of her upcoming trip to France. Without giving it any more thought, I forwarded the email to her.

Diane called at once. "You upset Karen with your email. Lots of people have told her not to go to France because it is anti-Israel. Alan must have told you to forward that email. Do you even know if it is true? Please don't send Karen any other emails without first clearing them with me."

CHAPTER EIGHTEEN

I covered my mouth with my hand and there was a lump in my throat. "I'm very sorry I upset Karen. That wasn't my intention. Alan wasn't involved at all. I acted without thinking, simply trying to connect with my niece. I have no political ax to grind."

Diane was still in full mama bear mode, protecting her cub. "Of course Alan had you send the note. He's the one who cares about politics." Diane loved me too much to accept that I was responsible. It was easier to shift the blame to Alan.

This conversation made me sad for a long time. I followed Diane's instructions and didn't communicate with Karen directly for many years. I wrote long, detailed, explanatory letters to Diane, but I never sent them. We continued to talk every Sunday morning without referencing the flare-up. Over time my hurt healed.

Five months after I sent the email, and while Karen was still in France, the antisemite Jean-Marie Le Pen was runner up in the 2002 presidential election. Also, during that time thugs attacked Jewish targets throughout the country. Later, Karen told me that Arab men she was hiking with made antisemitic slurs, and she at once told them she was Jewish. Not the response I would have suggested, but it worked for her. They stopped the insulting remarks and did not harm her. Karen thought she shocked them because they didn't know a Jew could have blonde hair and blue eyes.

While Karen was in France, her boyfriend Don got a leading role in a play. Karen surprised him by secretly flying home and showing up in the audience at the last performance. Once home, she decided to stay, unwilling to part from Don again.

When Noam was two, Jeffrey and Laurie moved to Wellington, Florida, to be near Laurie's mom. Tal, Laurie and Jeffrey's second son, was born a year later. Since the house prices there were reasonable, Diane and Ronnie bought one too, while keeping their Brooklyn co-op to use in the summer. The senior and junior Rothmans' homes were just minutes away from each other, and Diane and Ronnie loved the Florida lifestyle. Diane's grandsons kept her busy as she helped ferry them to school and activities. Also, she often had one of them sleep over. She told everyone, "Being a grandma is the best part of my life."

In Florida, Diane longed for some of the activities she loved in New York. One of these was her book club. By now she had recovered from her struggle with impostor syndrome and saw herself as a competent book club leader. She confidently organized a club from scratch and her enthusiasm attracted other members.

Diane's health continued to improve. Her desire to be there for her children and grandchildren motivated her to follow a complicated regime of pill taking, using alarms and ingenious pill boxes. While her heart strength (ejection factor) never improved, she lived a normal life. She played doubles tennis, walked, and traveled. Diane and Ronnie returned to their Brooklyn co-op each summer, where they enjoyed time with old friends and Diane played bridge, mah-jongg, and Scrabble at their beach club. She eagerly awaited seeing Karen married and reveling in Karen's children. She often spoke of how thankful she was for her many extra years of health.

Even though Karen and Don lived in New York, they were married at a Ritz Carlton in Florida. Diane and Ronnie were thankful to see Karen happy and embraced Don as a son. Everyone was relieved when Don made peace with his dad in time for him to come to their wedding. The bride and groom were adorable together. They put on a great show with a well-rehearsed dance routine worthy of professionals. After the wedding, they set off to Los Angeles to look for acting roles.

CHAPTER EIGHTEEN

But the marriage was in trouble from the start. Karen was lonely without her friends. Also, since Karen wanted to get pregnant, she went off her antidepressant to protect the fetus and fell into a depression. She and Don soon separated, and she returned to New York. Around that time, Judy got married. Inspired by Karen's wedding, she chose the Ritz Carlton hotel in San Francisco for her venue. Karen came to the wedding without Don. She wore a bright green silk dress, and at one point jumped onto the stage with the Texas Cowboys band to sing with them. Alan also surprised us by joining the band to sing our wedding song, "When I'm Sixty-Four." This was an enormous success since we and many of our guests were nearing sixty-four. Not long after Judy's wedding, the courts finalized Karen's divorce.

NINETEEN

Facing Fate

> "Don't waste your time worrying about all the terrible things that may happen to you. None of them will. It will be something else." —Cyrilla Weiss, my mother

The next year, Karen and Andrew found each other through Myspace. They first met in high school, where Andrew had a crush on Karen—like my dad with my mom. Nothing came of it back then. They both married other people and were now divorced with no children. Their reunion resulted in a storybook romance. Within a year, they were engaged.

Andrew was an associate in the law firm Squire, Sanders, and Dempsey. Karen taught third grade. They were thirty-five, and both wanted kids. They worried that Karen might not conceive because of her age. But a few months later, Karen was pregnant. The heaven-sent fetus prompted them to move the wedding date up a few months. Karen had a co-op in her parents' Brooklyn Heights building that the young couple were remodeling to create a nursery. They learned the baby would be male and planned

CHAPTER NINETEEN

to name him James Alex. My sister would soon have another grandson. Her son Jeffrey already had three boys. The third, Zev, was a surprise package who showed up five months before Karen's wedding. Everyone was euphoric.

The new wedding date of August 17, 2008, was close to my scheduled return from a scientific meeting abroad. I changed my flight to arrive in New York in time to attend the Aufruf, a religious ceremony before the wedding. The wedding, two days later, was a sweet ceremony in the Brooklyn Prospect Park Boat House. Those of us who knew the bride was three months pregnant detected a slight thickening at her waist as she flashed her brilliant smile and sashayed her sexy figure down that aisle.

For years, our daughter Judy's rant had been, "Forget about it. I am never getting married and never ever having children!" This had left us deeply saddened. We had also worried about our son Michael's lack of girlfriends and had mourned for the grandchildren we would never have.

Karen's wedding marked the end of this grief. Judy and her husband of two years confided they were newly pregnant, so the baby cousins would be just a few months apart in age. At thirty-two, Michael was there with a serious girlfriend, Susan, whom he was to marry.

Diane and Ronnie now had a home in Florida near Jeffrey and a co-op in the same building as their daughter in Brooklyn. Their plan was to spend half the year in each residence. Gone was the depression that dogged Karen for so many years. She was happy, secure, and in love. She, her new husband Andrew, and their parents were ecstatic.

Now I'll let Karen, Andrew, and Jeffrey talk to you directly as if on November 16, 2008.[1]

Karen and Andrew. The couple show off Karen's bump the day before her death. They were out celebrating on Andrew's thirty-seventh birthday.
FAMILY PHOTO

Karen:

I got married again. I'm thirty-six. This time everything is right. Andrew and I have so much in common. We love New York. With our jobs, we can afford to live here in style. We both had short marriages. When we were in high school at Packer, Andrew

CHAPTER NINETEEN

was already in love with me! He volunteered to manage the girls' basketball team just to be near me. I had no idea. Even our two upper-middle-class Jewish families mesh. We had a fabulous wedding and an enchanted honeymoon.

We wanted kids so much and worried I was too old. But we are already expecting!! We put together the apartment so our baby, James, will have a nursery. We have a crib, clothes, and everything. Jeffrey and Laurie gave us all the designer baby clothes that Zev outgrew. Judy is also expecting. The cousins will be just a few months apart.

Mom thought I'd never get it together. But here I am. Pregnant. Married to a great guy, Jewish and a lawyer. I'm teaching third-grade in a public school. We've been living in Park Slope in Andrew's place right near his family. Soon we will move into my apartment, in the same building as my parents, at 75 Henry Street. All of my dreams have come true.

Being pregnant is so much fun. I love how I look. The only problem is that I've been getting out of breath. I get winded climbing stairs, and I have a cough. My doctor says this is normal for a woman in my stage of pregnancy, and for a third-grade teacher surrounded by germs. She has no concerns. Everything is a go.

Andrew turned thirty-seven yesterday. I don't turn thirty-seven for another thirty-eight days. To celebrate, I took him to a matinee of "Speed the Plow," and then to dinner at the Aquavit in Manhattan. The food was amazing and the presentation artful. It was nice having Andrew's birthday fall on a Saturday.

Today we stayed in bed late, reading the Sunday *New York Times*. Then we grabbed brunch and took a walk in our Park Slope neighborhood. By about 2 p.m. I was so hungry. I *am* eating for two! We decided to try a Colombian restaurant, Cafe Bogota, that we keep passing. There were bright whimsical murals on the walls and spicy aji sauce on the tables. We sat at a table for two with cane-back chairs. Even though it was off hours, the restaurant was full. We just wanted a bite because we had reservations to eat out that evening with Andrew's mother, brother, and sister-in-law to celebrate his birthday. That's why we only ordered a few of the delicious down-home South American dishes.

When we finished eating, the waiter brought us a comment card along with the bill. I asked for a pen to rate the restaurant a ten. I started to write and . . .

Andrew:

Suddenly Karen stopped writing, lurched bolt upright, and stared at me with open eyes. I thought she was fooling around, but still I asked, "Karen, are you OK? Is anything wrong?" She collapsed forward onto the table. Her eyes remained open, but she didn't respond. As I held her, we slid onto the floor together. I screamed. She wasn't gasping for breath. She lay motionless. Police, EMT, and fire trucks arrived within minutes.

I heard someone shout, "Defibrillator!"
Questions came at me. "Was she taking any medication?"
"No."
"Any drugs?"
"No."
"Medical history that could explain?"
"No."

Methodist Hospital was nearby. Police drove me there behind the ambulance that took Karen. Meanwhile, someone used my cellphone to call my family. They were at the hospital when I arrived. Doctors told us they were working on Karen and delivered James by emergency cesarean section.

But soon they returned, this time with the hospital chaplain. "We could not revive your wife. She never recovered consciousness. She is dead. We are sorry for your loss."

I collapsed on the floor in intense physical pain. How can she be dead? We still have our whole lives ahead of us. Young, healthy people don't just die!

CHAPTER NINETEEN

But in my gut, I knew it was true. I had been with her. I saw her lifeless, open eyes. I knew she died instantly. There was no pain. No time for regret. No realization of death to come.

Later they came back to complete my misery. Baby James had died as well. They were sorry they weren't able to save him. They took me to see Karen's body and to hold James for the first and last time.

Jeffrey:

Laurie and I and the boys were eating dinner when I answered a call from Andrew's sister-in-law. It was very odd to get a call from her.

"Jeffrey," I stopped feeding Zev and focused, "something happened to Karen. She's in Methodist Hospital. She passed out in a restaurant. Bystanders there called me. Does Karen have a medical history of seizures?"

"No, she's always been healthy. Is the baby OK?"

"I don't know. I'm heading to the hospital now. I'll call you back when I learn more."

Laurie had had a premonition of a disaster the night before. She had been up crying and looking at pictures of Karen. Now we tried to make sense of the phone call as we prayed that the baby was safe and waited for more news.

I picked up the phone in mid-ring. Andrew's sister-in-law was on again. This time she was hysterical. Between sobs she uttered the incredible. "Karen died. I'm so so sorry."

"That can't be. How do you know?"

"I'm at the hospital. The doctors just pronounced her dead."

I slapped my head with my hand as I jumped out of my chair. "What? That can't be right. Karen is young and healthy. You must be wrong."

Laurie saw my distress and stood beside me. "Jeffrey, what's wrong? Is Karen's baby OK?"

I shook my head no and waved my hands repeatedly as if to say, "It's finished, it's finished."

The voice on the phone kept insisting. "I heard it from the doctors. I'm very sorry, but it's true. Karen and the baby both died. Doctors couldn't revive them. The cause of death is unknown. I am so, so sorry. Do you want me to call and tell your parents?"

"No, no, no! That's my job. I will do that. That's mine."

Laurie and I drove to my folks, who lived near us in Wellington. As soon as they answered the door, they could tell something was very wrong. I knew I had to be the one to tell them, but this was the hardest thing I ever had to do.

We walked past the cheerful dining room area, with a modern glass table, bright turquoise leather chairs, and a rectangular crystal chandelier, and arrived in the small living room.

"Are the boys OK?" my mother asked as her hands trembled.

"My kids are fine. I want you to both sit down."

The couch faced a large painting of magazine clippings of famous past events. To the right were glass sliding doors leading to the patio and backyard. Toys for their grandchildren lay scattered about. My news would soon annihilate this domestic tranquility.

"Jeffrey, you're scaring us. What's wrong? Did something happen to Karen's baby? Tell us what's wrong."

"Please just sit down." When they followed my instructions, I told them. "Karen died a couple of hours ago, and the baby, too. She collapsed in a restaurant while out with Andrew."

Mom and Dad jumped up and embraced each other as they screamed in anger and disbelief. Mom started wailing and writhing on the floor. Dad paced while hitting his head, and shouting, "I thought I'd have more time. I thought I'd have more time . . ."

Then my mother leaped up, grabbed her purse, and announced, "I'm going to the airport. I want to see my baby girl right now. I want to see my baby tonight."

We caught a flight that night. When we got to the morgue, Karen was cold from being in refrigerator storage. A pathologist

CHAPTER NINETEEN

would perform an autopsy in the morning. Her throat was still plugged from the rescue attempt. Otherwise, she looked like she was sleeping. Mom didn't want to see her with her mouth blocked. She focused on her feet, rubbing them to say goodbye.

Jeffrey called me that night and I caught a flight the next morning. Alan followed a day later. I stuck to my sister as much as possible, before, during, and after the funeral.

With this kind of death, an autopsy was mandatory. Also, the family desperately wanted answers. The ruling was no foul play, no poison, a natural death. But how, why? No aneurysm. Her heart seemed fine, but the New York City Coroner sent it for further study. These results and their shocking revelation wouldn't be back for weeks. Meanwhile, we all feared for Judy, four months pregnant. She told her doctor what happened to her cousin, and he checked her out with no negative findings. That was small comfort, since Karen's doctor said the same thing to her when she complained of being out of breath and having a cough. We learned and worried about QT syndrome, which causes sudden heart arrhythmia and death and runs in families.

There was much discussion of whether to follow the Jewish tradition and bury Karen in a white shroud, as preferred by Andrew, or, as Diane insisted, in her wedding dress. It is irrational how something so unimportant can loom so large. Since Diane and Andrew couldn't control Karen's death, it became critical to control what she wore for eternity. Andrew argued that the wedding dress represented the happiest day in Karen's life and shouldn't be buried with her. He was adamant that the memory of that day and something so joyful as the dress should not be spoiled by the sadness of her death. Diane relented and the undertaker wrapped Karen in a simple white shroud.

She lay with baby James in her arms, next to my parents. My dad—Karen's grandfather—died twenty-eight years ahead of Karen to the day, on November 16, 1980. His father, David Weissman, was born on this same day in 1888.

I signed over the last burial plot to Andrew, so he could be laid to rest near Karen and James when his time came. We realized Andrew would likely find a new love and not want to use it, but for the moment, it comforted him. Also, Diane liked to think of Andrew joining Karen someday.

I slept on the couch in Diane and Ronnie's living room that week. My hair went wild because I didn't have my brush hair dryer. I kept that hairstyle for years afterward, as a reminder of Karen.

The funeral was massive and emotional. Andrew had the strength to read the poem known as "Stop all the clocks" by W. H. Auden.[2] When he read it, he switched the masculine pronouns to feminine to make the poem reflect his loss.

At the end of the service, ushers guided us into waiting limousines to take us to the cemetery. At the cemetery, other mourners had to physically support Andrew. Jeffrey, with the aid of family and friends, shoveled all the dirt necessary to bury Karen. He didn't want strangers to do this.

After the funeral, visitors came to console the mourners at the shiva. Diane's friends surrounded her with love. One in particular had lost an eleven-year-old son who was hit by a car thirty years earlier. When consoling Diane, she was finally able to cry for her son for the first time.

At one point, Diane complained that certain neighbors just came to gawk and socialize. "Do you want me to ask them to leave?" I asked.

"Yes, if we can do that."

Finally, I could help her with something. "Thank you so much for dropping by to pay your respects. This is a grim time

CHAPTER NINETEEN

for the family. They only want loved ones and close friends to surround them now. I'm sure you understand."

"Are you asking us to leave?"

"Yes, please. I'd appreciate it."

Besides the full house of visitors, there were many telephone calls during the shiva. I listened as Jeffrey talked to Karen's first husband, Don. At first, he assumed it was suicide.

"No," Jeffrey explained, "she was remarried, pregnant, floating on air, and then, just like that, she left us."

Jeffrey warned his boys that Grandma would never be the same. "Don't expect her to play with you or even to smile ever again."

But Diane was stronger than we thought. When holding baby Zev, she announced that if not for that little boy she would have definitely jumped off the high-rise terrace. "Without you, Zevie, I would have jumped for sure."

"What about me?" complained Ronnie.

"Don't worry—I would have taken you with me."

Jeffrey corrected himself, "My mistake. Yup. Same old Grandma."

Diane and Ronnie even returned to traveling. Diane felt a strange comfort in knowing that nothing bad could ever happen to her baby girl again. And it was a consolation that Karen had found happiness before her death.

My daughter, Judy, gave birth five months after Karen and James died. Judy and Ben wanted to name their baby after Karen, so they chose Aaron because it rhymes with Karen. Years later, Jeffrey told Judy that he could see some of Karen's free spirit in Aaron.

TWENTY

Hunting the Killer

"Each player must accept the cards life deals him or her. But once they are in hand, he or she alone must decide how to play the cards in order to win the game." —Voltaire

The medical examiner's report we received in the wake of the tests on my niece Karen's heart sent shock waves through the family. The coroner ruled that Karen died from dilated cardiomyopathy (DCM), the same disease that afflicted her mother, my sister Diane. This meant that Diane's illness was genetic and wasn't a complication of a viral infection, as doctors formerly assumed. Instead, Diane transmitted a deadly gene to Karen. We had to find the mutation before God's cruel game of Russian roulette picked off more members of the family.

Our hunt for the mutation began when Diane saw a genetic heart disease expert at the nearby University of Miami. Dr. Ray Hershberger explained that scientists knew of about ten genes that could cause DCM when mutated. Mutations in these genes were dominant, so inheriting a mutant copy from one parent could cause disease, even in the presence of a good copy from

CHAPTER TWENTY

the other parent. One of our parents may have carried a good and bad copy of the gene and could have given Diane the bad copy. This parent would also have had a 50 percent chance of transmitting the bad copy to me. If I had inherited the bad gene, did I pass it to my children, Michael or Judy? Did Judy then pass it to her unborn baby?

Alternatively, the mutation may have first occurred in Diane. Then I would not be at risk but there would still be a 50 percent chance that Diane gave her son the bad copy. If Jeffrey inherited it, did he pass it to any of his boys? Who would suddenly die next? When would this curse end?

Doctors recommended that the adults in the family get yearly echocardiograms to detect and treat any heart abnormality early. Alan didn't want me to frighten our children with this recommendation. He worried it could discourage them from having their own children. He considered Karen's death a fluke and wanted us to move on. We compromised. I would tell our children of the recommendation once but was not to remind or question them about it in subsequent years.

Meanwhile, Dr. Hershberger and his team sent Diane's DNA to Harvard to analyze it with the fledgling DCM genetic screening test then available. Jeffrey took part in these meetings. He was eager to have his DNA tested as soon as they found the mutation that caused his mother's disease. Within six months of Karen's death, we had the results. Diane didn't have mutations in any of the genes tested.

Next, I consulted Dr. Elizabeth McNally, a genetic heart disease specialist at the University of Chicago. I heard about her through the Chicago yeast genetics community to which both her husband and I belonged. During the next few years, she repeatedly ordered tests of Diane's DNA for newly discovered genes suspected of being able to cause sudden death and DCM when mutated. She didn't find anything. This left Diane and me frantic to protect our children and grandchildren.

My sister, Diane, dancing at my son's wedding. This was in 2010, two years after Karen's death. FAMILY PHOTO

CHAPTER TWENTY

It cost a billion dollars in 2003 for the Human Genome Project to complete the first sequence of a human genome. By 2007 innovative technology caused the price to plummet, and by 2010, the cost of sequencing an individual human genome was about $50,000. This made it practical for Dr. McNally's research grant to invest in sequencing several of her patients' genomes with the goal of finding new heart disease genes. My sister's genome was among those she analyzed. By 2024, it cost under $500 to sequence an individual human genome. Alas, many doctors are not aware of this dramatic price decline and avoid DNA testing, thinking it is exorbitantly expensive.

Since many harmless mutations are present in everyone's DNA, finding the specific mutation responsible for my sister's disease wasn't easy. Lisa, the genetic counselor in the group, called to tell me that by focusing on genes likely to affect the heart, Dr. McNally had pinpointed a smoking gun. The suspected mutation was in a gene that coded for a protein that was a part of skeletal and heart muscles. To help prove their hypothesis that the mutation affected muscle function, she asked Diane and me to send fresh blood samples. They wanted to reprogram our blood cells into muscle to test if the mutation compromised the muscle's function. Diane and I were happy to comply. We hoped the end was in sight. For now, they didn't want to tell us the name of the suspected gene because the evidence was not conclusive.

About this time Alan and I moved from Chicago to Reno, Nevada, just a four-hour drive from our kids and their families in the San Francisco Bay area. I took my grants with me and secured a soft money (grant-funded) research faculty position at the University of Nevada, while Alan continued his home-based financial advisory business. When we got to Reno and I wasn't Dr. McNally's patient anymore, I emailed Lisa, the genetic counselor, over the next two years to find out about their results. Since I didn't get any responses, I assumed progress had stalled.

Diane developed severe back pain in the last year of her life. Doctors excluded the possibility of surgery, warning that general anesthesia was too dangerous for her. They put her on heavy pain medication instead. She took to her bed and rejected visitors. We had only brief phone conversations during that year.

Then magically, the pain lifted. Diane started making plans to attend a family wedding in Iceland in a few months, where she would see the Northern Lights on New Year's Eve 2017. She and Ronnie even had plane tickets. But once she got out of bed, the heart failure was back. The doctor prescribed a pacemaker. After a simple implantation through a vein in her arm failed, Diane left me an upbeat phone message that she would have surgery the next morning. I didn't find this recording until my nephew called to tell me Diane was in a coma following the anesthesia. The next day she woke up, but she was abnormally agitated and not herself.

She died the following morning, April 21, 2016, the day before Passover, eighteen years after her diagnosis, and seven and a half years after Karen's death. She was seventy-three. Following the Jewish custom, I light a Yahrzeit candle on the anniversary of her death to honor her memory. The white wax of these candles fills a glass and burns for 24 hours. I like to imagine that her soul and warm love are near, signaling to me through the yellow flickering Yahrzeit flames. It's hard not having her here.

After Diane died, I wrote to Lisa to ask about the mutation again, but this time I copied Dr. McNally on the email.

> I just returned from my sister Diane Rothman's funeral. She finally succumbed to heart failure. Suffering this loss and being with the family propelled me to contact you again and ask if you have made any progress in identifying the gene that caused Diane's death. What happened with the suspected gene you were following up on in 2014 with the muscle cells you made from Diane's and my blood? What, if anything, do you

CHAPTER TWENTY

anticipate doing on this case in the future? Also, since it has been several years since our last genetic test, do you think there are now new known genes that could be checked with Diane's blood? Diane has a son and three grandsons, and I now have three grandchildren so I am anxious to protect them if at all possible.

Beth McNally replied swiftly.

So sorry to hear about your sister Diane! As you can see by the email, we've moved to Northwestern (Lisa as well, copied above, using her married name). We found a putative splice site variant in a gene not previously linked to cardiomyopathy in Diane. At the time, it was the only link to cardiomyopathy for this gene (FLNC), but now there have been a few more reports potentially linking this to cardiomyopathy so that makes it more likely that this may be a real association. Lisa can set up a time to review with you. Has your heart continued to do well? Again, so sorry to learn of Diane's passing.

I hadn't received replies to my earlier emails because the University of Chicago didn't forward them to Lisa at Northwestern. From my phone conversation with Lisa I learned that Dr. McNally's group had described a mutation they thought caused my sister's disease in a paper they published two years earlier.[1] The mutation was in an intron in the FLNC gene. Human genes contain protein coding regions known as exons, as well as regions called introns that are scattered throughout the coding sequence that can be thought of as junk. To make the correct protein, introns are spliced out of the mRNA using special splice site sequences at the exon/intron boundaries. Many genetic diseases occur when there is a mutation at one of the exon/intron boundaries because this leads to improper splicing.[2] This is what my sister's mutation did when it changed the last letter of an intron, a G, into a C. The failure to correctly splice out an intron messes-up the mRNA sequence and frequently leads to a premature stop codon identical to the protein synthesis termination/stop codons I studied in yeast. Cells recognize and degrade mRNA with premature stop codons using a process called nonsense mediated mRNA decay. Indeed, cells with our mutation have only about half the normal level of FLNC mRNA, presumably made from the good copy of the gene without a premature stop codon.[3]

I Googled the mutant sequence and found that another DCM patient had the same exact mutation.[4] The authors of this paper published their findings at about the same time as Dr. McNally did, but the two groups hadn't seen each other's work. Coincidentally, the senior author of the paper, Dr. Frederick Roth, was the son of a scientist I knew. Like me, his father, John Roth, worked on mutations that affect the recognition of termination codons in yeast. I visited John at the University of Utah when I gave a talk there many years ago. He had a huge, white, furry dog in his office which I mistook for an ottoman! Frederick and I also had other colleagues in common. We had a fruitful email exchange. I hoped we could find out if my family and the patient described in his paper were relatives.

Finding the identical sequence change in the two patients made us wonder if they were related. Unfortunately, the other patient had checked off "do not contact me" on a form, which prevented us from learning anything about him. All we had was his gene sequence and a diagnosis of DCM with an ejection factor of 19 percent from an earlier doctor. Nonetheless, this finding bolstered the case that the suspected mutation indeed caused my sister's DCM.

I emailed Beth and Lisa:

> Learning from your paper that I do not have the mutation changed my life. No longer do I panic every time I get a little out of breath. Instead, I congratulate myself on getting exercise. I had come to expect that I might die soon while knowing that I might not. Now I am expecting to live a long time, knowing that I might not. This change in perspective makes a dramatic difference! Similarly, your work has lifted a great weight from my daughter about her and her boys' futures.
>
> Most relevant for genetic counseling was my son's response. Despite my clear instructions and his visit to the Stanford University genetic cardiologist you recommended, he wasn't getting his heart checked be-

cause his GP told him it was unnecessary. She told him that since I was healthy, he couldn't have inherited a bad gene! [This doctor didn't know that I could have had the disease-causing mutation without ever getting the disease. This is because the mutation's penetrance and expressivity, which is the chance of it actually being expressed and causing disease, is less than 100 percent. Importantly, I could have passed the mutation to my son where it could have caused disease.] My son hardly remembered anything in terms of symptoms to watch out for either. Clearly another important reason for genetic testing is that learning you have a pathogeneic mutation would capture your attention and encourage you to keep getting your heart checked.

To share the news with Diane's son I triumphantly texted, "We've got it. We know the mutation! Your doctor can now order a test for you. Check your email for details and call me."

"Hurrah!! I'll get on it ASAP," he texted right back.

But then a later message that day said, "Laurie doesn't want me to get tested. I will have to bow to her wishes and fears. This is too distressing for her."

What? We finally have the gene and now his wife doesn't want to find out if he has the mutation? My unspoken contract with Diane was to protect Jeffrey and the boys, then eight, twelve, and fifteen, from this scourge. How can I do that if Jeffrey doesn't get tested?

As a physician, Laurie felt they were already doing everything they could to avoid Karen's fate. They were on heart-healthy diets. Jeffrey was getting regular echocardiograms, and when the boys were older, they would get them too. In Laurie's view there wasn't anything more to do even if they knew Jeffrey had the mutation. She worried that learning that her husband or one or more of her sons had the deadly change would be too upsetting. She didn't want to know.

I had to respect their decision. I had hoped that Jeffrey would discover he hadn't inherited the mutation, ending the dread that he and/or one of his sons would suddenly die. In contrast, Laurie feared that if they found they had the mutation it would be too devastating. In my view, knowing they carried the mutation, which might or might not ever cause disease, wouldn't be too different from their current state of knowing they *might* have the mutation.

Meanwhile, I asked Dr. McNally if she would sequence the suspect gene in Karen's DNA if I could get a postmortem sample sent to her. This was to confirm that one of Karen's FLNC genes contained the mutation as expected if we were correct and the mutation caused the disease. Dr. McNally agreed.

As Karen's aunt, I had the power to ask the coroner to send her tissue sample to Dr. McNally. I called the pathologist who did Karen's autopsy, Dr. Barbara Sampson. She well remembered the case. The sudden death of a beautiful, healthy thirty-six-year-old pregnant woman stood out even among the many autopsies she performed. In the eight years since Karen's death, Dr. Sampson had become the Chief Medical Examiner of the New York City coroner's office, the first female chief in their history. Her office had set up a research project on sudden death so they were also interested in the new gene.

After Dr. McNally determined that Karen's DNA indeed contained the mutation, she recommended that I confirm her laboratory sequencing of my gene with clinical testing. "This is necessary," she explained, "because research sequencing doesn't use all the controls required of clinical labs." I sent the genetic testing company GeneDx my saliva sample, and the New York City coroner sent a new sample of Karen's tissue to them to use as a control.

When these clinical results were available, Lisa called to tell me that GeneDx confirmed the research lab findings. I didn't inherit the disease mutation. Since I didn't have a copy of the mutated gene, I couldn't have passed the mutation to my children, or they to their children. We had indeed won the lottery. We were truly free of the danger I had feared for the last nine years. Why had fate spared me while dooming my sister?

CHAPTER TWENTY

During this conversation, I asked Lisa if clinical DNA sequencing labs listed their findings somewhere to inform doctors and other researchers of their results. That's when she told me about the newly established ClinVar[5] website that did just that.

When I looked up the mutation that Diane and Karen had on ClinVar, it amazed me to find two more families with the same exact mutation.[6] In the family of a woman with the mutation who had a heart rhythm problem, the mutation had caused the sudden deaths of her son at seventeen, her brother at thirty-two, and her grandfather at forty-five. In the other family, the use of a defibrillator averted the sudden death of a forty-four-year-old male with the mutation.

The data in the paper showed that most, but not all, who inherited the mutation in these families had either DCM, irregular heartbeat, or sudden death. In contrast, no family members without the mutation suffered from these problems.

When I contacted the paper's authors, they treated me like a colleague even though I told them I was a *yeast* geneticist. They had also worked with yeast and appreciated the contribution that the yeast model system made to human genetics.

Indeed, in 1996, yeast was the first organism, after a few viruses and bacteria, to have its entire DNA sequenced.[7] Unlike viruses and bacteria, and like human cells, yeast have their DNA packaged into chromosomes that are in a membrane-bound nucleus inside the cell. The methods used and lessons learned in sequencing the yeast genome were crucial in guiding the much larger Human Genome Sequencing Project.[8]

During our email exchanges, one of the authors, Dr. Michael Arad of Tel Aviv University, shared unpublished information. He revealed that both of the families he treated with our mutation were of Ashkenazi Jewish descent. This was telling because my family was also Ashkenazi Jewish. All Ashkenazi Jews originated from a population of less than five hundred in Eastern Europe whose genetic origins were equally split between the Levant and Europe so we were indeed distantly related.[9]

I looked this mutation up in the Inflammatory Bowel Disease Exomes browser. This publicly available database shares the DNA sequence analyses of all the protein-coding (exome and intron boundary) regions of 5,685 Ashkenazi Jews, 7,240 non-Finnish Europeans, and 10,626 Finnish. Remarkably, the mutation in my family occurred in one out of 800 Ashkenazi Jews but was not present in the other ethnic populations.[10] The high frequency of the mutation specifically among Ashkenazis suggested that the "founder" mutation first appeared in one member of a small Ashkenazi group that grew through inbreeding. Antisemitic assaults have periodically decimated the Ashkenazi Jewish population. All ten million Ashkenazis alive today are descendants of a small surviving group that lived about five hundred years ago. Clearly, the mutation has been around for a long time and didn't first occur in my sister. Other examples of founder mutations in a variety of different genes exist in the Ashkenazi population as well as in other ethnic groups. Founder mutations are always established by inbreeding.[11]

Dr. Arad and his co-authors worked with many families from all over the world and found twenty-three different mutations that caused truncations of the same protein (FLNC) that my sister's mutation effected.[12] Of the twenty-three different disease causing FLNC mutations they found, only our mutation was associated with Ashkenazi Jews. Alarmingly, some of their FLNC patients had normal echocardiograms but still died suddenly. Hence, my nephew's wife's reliance on echocardiograms to protect her family was not sufficient. Rather, the authors suggested that patients with the mutations also undergo periodic Holter monitoring and cardiac MRI evaluation. If these tests detected irregular heartbeat or fibrosis, doctors could recommend the patient get a defibrillator implanted in their chest to thwart sudden death. This is important because of eighty-eight patients with similar

CHAPTER TWENTY

mutations (FLNC truncations), 15 percent died suddenly. Doctors would treat symptoms of heart failure, dilated cardiomyopathy (DCM), or heart damage with guideline-directed medical therapy. This includes use of medications such as angiotensin converting enzyme (ACE) inhibitors, beta-blockers, Entresto, diuretics, and/or angiotensin II receptor blockers. These therapeutics forestall added heart damage, improve heart function, and enhance the quality of life. If they do not successfully combat heart failure, heart transplants can provide a cure for decades.

This information convinced my nephew's wife to let her husband get his gene sequenced. The value was that if he hadn't inherited the mutation, they could avoid frightening their boys with frequent Holter and MRI tests. We would soon know if my nephew and his sons carried the family curse. If they did, we could use modern medicine to protect them.

The tragic story of my uncle's accidental death haunted me for decades. Learning of the accident also affected my cousin Marty's life. We talked about it together for the first time in November 2017, at Marty's 50th wedding anniversary celebration. The party took place at a Fireman's Hall in Red Lion, Pennsylvania, near where Dad grew up. Family photos hung on the far wall of the large room. Marty's wife cooked up a storm for the buffet at this casual party, and one of their adult sons played in the live band.

It was difficult to talk while the band was playing, but during their break we had a chance to compare versions of the tragic family story. It was hard to piece together what actually happened. I wondered why my grandmother Marion wasn't watching her son. I also didn't understand how Marion could continue as a dressmaker using mirrors like the one that killed her son, Eugene, since she had other young children. Marty's and my parents had died years ago so we couldn't check details with them.

While we spoke, the grandchildren enjoyed a wooden ring-toss game. Marty defended Grandma Marion, explaining that my version of the story differed from what he had heard. Sipping a beer, Marty suggested Eugene was born after, rather than before, my father. It made sense for Eugene to be the middle brother since my father was ten years older than Marty's father, Cyrus. Marty also recalled hearing that our grandparents weren't home when the accident occurred. He believed my father, the oldest brother, was babysitting when the mirror fell.

My heart sank. Could this be true? Was it possible that the mirror fell on and killed Eugene while my father was in charge? Did Dad shoulder this guilty secret all those years? These thoughts plagued me for the rest of the weekend and during my return flight to Reno.

After I arrived home, I searched a genealogy website to investigate Marty's story. The father I knew wouldn't have lied about the accident or concealed his role. The first thing I found was Eugene's death certificate. Since I knew the cause of death, I focused on learning that Eugene died at four and a half on August 10, 1916. Next, birth records showed that on that day my father Norman was two, and Cyrus wasn't born yet. I also found a photograph showing that Eugene was my father's older brother. That was a relief. Dad couldn't have babysat when he was only two! Although Eugene didn't die before Dad was born, Eugene was indeed his older brother. Also, my father wouldn't have remembered Eugene because he was so young at the time of the accident, so his statement that he didn't know Eugene was the truth. Three years later, I was to discover a grim secret associated with the accident that my father never knew.

TWENTY-ONE

The Dressmaker's Secret

"Forewarned is forearmed." —Old Proverb

My family endured sudden deaths for centuries. Now, for the first time, there was something we could do about it. Lisa and Dr. McNally wrote an official "Dear Family" letter from the Center for Genetic Medicine at Northwestern Medical School for me to send to relatives explaining the dangers of the deadly mutation and offering to facilitate their genetic testing (appendix 1 is the text of the letter).

Since I didn't have my parents' DNA and there wasn't a clear family history, so I didn't know if the mutation came from my mother or my father. As a result, I warned both sides of my family. In the process, I found many relatives I didn't know.

I contacted all first and second cousins, who, respectively, had a 12 percent and 6 percent chance of having the mutation. Those who knew of my niece Karen's death were at once interested. Those whom I didn't know and who didn't know Karen were sometimes skeptical about the relevance of my information and at first thought I might be part of a scam.

CHAPTER TWENTY-ONE

It took a while for me to refine my email and telephone pitch. After establishing our genetic relationship and discussing family stories to gain trust, I informed them that I called (or wrote) to warn them of a mutation in the family. Without further preamble, I told them that they were at X percent risk of having a mutation that could cause heart failure and sudden death, but that there were treatments to prevent these dire outcomes and that a simple relatively inexpensive test could determine if they had the mutation. Also, I explained that if they did have the mutation there was a 50 percent chance that they passed it on to each child. I then told them about the clinical treatments, penetrance and expressivity (the chance they would get sick if they had the mutation), and the Ashkenazi connection. Finally, I told them exactly how to get tested. I contacted the oldest generation and explained that unless they had the mutation, their children couldn't inherit it. This was a relief for families who knew about Karen's death and worried about their own and their children's danger.

I was already in communication with my five first cousins. I easily found the second cousins on my maternal grandmother's side because that side of the family descends from a Tzaddik and has an extensive family tree. The title "Tzaddik" means a righteous man and spiritual leader. Hasidic Jewish communities gave their Tzaddiks absolute respect. They thought a Tzaddik was an intermediary between God and man. Joseph Moses Abraham, a Lithuanian scholar and Rabbi known as the Tzaddik of Lazday (1754–1824), was my Grandma Jennie's great-great-grandfather.

Grandma Jennie was born in Russia and was the second youngest of six sisters and a brother. When her brother came to the United States, he worked at night cutting out fabric and completed medical school during the day. He was the second oldest and helped pay for the rest of the family to immigrate. Jennie always looked up to him and his daughters, who married doctors, but over the years we lost touch with them.

The family moved into a second-floor brownstone on the Lower East Side of Manhattan. When the mailman announced the mail each day, he had trouble pronouncing their name, Sukashefsky, so he suggested they change it to something easier. They obliged and chose Hymanson from their father's first name.

I knew the descendants of all but the two oldest of these Hymansons. Many lived in the New York City area and kept up with my sister until her death. They welcomed my calls and emails, and we envisioned a renewed acquaintance.

The offspring of the two oldest Hymanson siblings, Rose and Abraham, were unknown to me, but it was easy to find their descendants and they accepted me as a relative. My mother had told me about Rose's family. Rose's oldest son, Ben, caught TB and had to move to a sanitarium out West. Then, his younger brother, Harry, died tragically and unexpectedly, leaving behind a young family. According to Mom, the family very much loved and respected Harry. Since Ben lived so far away, they thought they could hide the heartbreaking news from him. When Harry's letters stopped coming, Ben asked why. The family fabricated stories but Ben guessed and concluded, "If you can stand it, I can stand it." Sadly, Rose's other two children also predeceased her.

To find the remaining cousins I became an amateur genealogist. I used the Ancestry.com website, Facebook, newspaper archives, LinkedIn, white pages, real estate records, maiden names in college yearbooks, and Google.

Despite extensive efforts, I couldn't find relations of my maternal grandfather, Simon Zuchtmann. I unexpectedly found relatives of my paternal grandfather, David Weiss, when 23andMe matched David's brother's grandson and me as second cousins. This cousin then helped me find more relatives in his branch. These newfound cousins and I had a surprising amount in common. There were scientists, engineers, doctors, and artists in the family.

My father's mother, Marion, was one of seven surviving siblings. I knew of Marion's older brothers, Harry and Joseph,

CHAPTER TWENTY-ONE

because they owned "The Millner Brothers" print shop where my father worked while in college. I also remember meeting Marion's younger siblings, Normie and Sylvia, who were still children when their mother, Sarah, died. Marion took her sister Sylvia in while her brother Joseph took in Normie. My father was Normie's nephew although Dad was a year or two older. They liked to joke about that. I remember Normie and his wife, Alice, who taught in my mother's school and was her friend.

I had a lot of trouble finding Harry's offspring. His surviving nephew told me he recalled a cousin named Mona. I learned from Ancestry.com that a Mona Millner married an Edward Morris. The breakthrough came when I simply used Google to search for "Mona Millner Morris," and got a hit showing her attendance at a Brooklyn College alumni event. I wrote to Brooklyn College and got her last known address. A search for that address on the web uncovered the owner, who I soon discovered was Harry's granddaughter. From her, I learned of Mona's sister, Millicent, who had been a high school art teacher. Tragically, she died young of cancer. I found her son on Facebook and sent him a private message. He was delighted to hear from a relative, and we spoke on the phone the next day. I followed up with the following email.

March 10, 2018

Dear Dxxx,

So nice to talk to you today. Here is the promised email and attachments about the sudden death gene in our family. Again, I am Susan Weiss Liebman, granddaughter of Marion Millner who was sister of Harry Millner. This makes us second cousins! My father, Norman Harold Weiss, worked in the Millner Brother Print shop in high school and college. He died in 1980 of a sudden heart attack at age 66. The gene mutation in my family has just been discovered and is not part of any general screening test. However, it is easy to get tested for this new mutation now that we know what it is. The information on how to get tested is in the attached "Letter to the Family" [see appendix 1] and below. The other attachment is a scientific paper about our mutation and other mutations like it.[1] In summary, carriers of the gene should be checked for the heart problems described below and if found, a defibrillator can be implanted to protect from sudden death. Here is what the author of the attached paper wrote to me when I asked if implanted defibrillators have saved any carriers. "We have information on three affected carriers in different families who received appropriate ICD shocks from devices that had been implanted as primary prevention. The ages at these shocks were 33, 38 and 51 years old, respectively. Two of these patients received more than one appropriate shock" (2 and 4!! times).

As I mentioned, my niece, Karen, died suddenly and unexpectedly 9 years ago at the age of 36, and we now know that the cause was a mutation in the FLNC gene. The specific mutation was (c.3791-1G>C). Karen's mother, my sister, Diane had the same mutation, and it was the cause of her dilated cardiomyopathy and heart failure. She lived, traveled, and enjoyed life with medication until age 73 when the medicine could no longer prevent her heart failure. Recently the exact same mutation (c.3791-1G>C IVS21-1 G>C) was found in two other unrelated families that are descendants of Ashkenazi Jews. The mutation has not been found in any other ethnic groups. It thus seems that this mutation has been around in our family a long time, from a "founder" of the Ashkenazi population. One important thing to keep in mind about the mutation is that it is possible to have the mutation and pass it on to your children even if you never have any heart disease symptoms. This would be an absence of "penetrance." Another important thing is that if you have the mutation, it can cause irregular heartbeat that is unnoticed by you and can lead to sudden death even at a very early age (earliest I am aware of is 17 in attached paper[2]). The paper talks about our specific mutation FLNC c.3791-1G>C as well as other mutations that cause a similar change to the FLNC protein and can all cause sudden death. According to this paper, while both men and women are affected, men are generally affected more, and more often women can have the mutation without getting sick. However, they still have a 50% chance of passing the gene on to each child.

We do not know if the mutation in our family came from my mother or my father's side, but we think it is most likely from my father through his mother and thus related to you. We think this because my father died suddenly at 66 and his mother, Marion, died suddenly in her sleep at age 60. If indeed Marion Millner had this mutation, then Harry had a 50% chance of having it; your mother 25% chance; you 12.5% chance. Since Harry was not sick the chances are even less. If you get tested and do not have the mutation, then none of your children or grandchildren can ever inherit it from you and you will know that they are not in danger (at least from this). In the unlikely chance that you find that you have the mutation I would very much appreciate your sharing that information with me confidentially so I will know that my mother's side of the family doesn't need to be tested and that I do not need to search for my paternal grandfather's relatives. Likewise, if I learn that the gene has been found on my mother's side, I will inform you.

I got involved in this because of our terrible loss and also because I am a geneticist. I am a yeast, not human, geneticist, but I can still read the papers and understand about mutation, protein, and genetics. I am happy to talk with you, your wife, or your doctor about this if you have questions now or in the future.

The attached "Dear Family" letter is from my doctor's office and the attached scientific paper is about the mutations. There is a supplement to the paper with the family pedigrees I can send if you are interested. The family letter speaks about both HCM [hypertrophic cardiomyopathy] and DCM [dilated cardiomyopathy] but actually our mutation seems to specifically cause DCM not HCM. Also, while it is true that men and women inherit the mutation equally, generally often men who have the mutation show heart symptom by age 40-50 while women sometimes never show symptoms (although it killed my niece and sister).

All the best to you and your family.
Cousin Sue

P.S. Below are excerpts from the letter I wrote to my nephew that may tell you a bit more.[3] Also, FYI, here is a link to my art http://sue176.wixsite.com/sues-art

Please respond so I know you received this! I hope we can keep in touch. I will work on writing up information on all the Millner family and descendants that I have learned about so we can all get in contact with lost family.

It surprised me to find two more Millner siblings, Ella and Leah, that I didn't know about when an unknown second cousin of mine appeared in 23andMe. Through this connection, I learned of the long-ago sudden death of Ella's thirty-two-year-old grandson while snorkeling on vacation in the Bahamas. The autopsy ruled the cause of death was heart disease. He left behind a young son and daughter. I spoke with his widow and son. These must have been the relatives that Uncle Cyrus referred to when I asked if there were any other family tragedies like Karen's sudden death. He said there was one but insisted it was a fluke and shouldn't be of concern.

Eventually, I realized that my father, not my mother, was very likely the source of the mutation in our family. The decisive recollection was Mom's warning to avoid short-distance running because Dad's intensive participation in that sport had enlarged his heart. Since DCM, but not exercise, causes a large heart, Dad's problem must have come from the mutation, although we can't confirm this without a sample of his DNA. It is now possible to bank samples of your DNA easily and inexpensively so they will be available to future generations (see appendix 2).

With my sister gone, I communicated more with her son, her husband, and his new wife. I thought Diane had told me everything. This was not true.

Yes, we missed some years before Dad died when we only communicated through Mom and Dad—the operators. But when they were gone, we talked for hours every Sunday. First, we mourned together. Then it was a give and take. She tried hard to listen. I tried hard to speak. She gave me advice. She showed she cared. I heard many stories. She loved to embellish.

After . . . well, after . . . we kept talking, but then everything was different. After . . . I knew my role was to listen, to love, to support. She told about nasty bridge players, changes in bridge partners, unfair homeowner association rules that funded Christmas but not Hanukkah parties, new friends and grandchildren, grandchildren, grandchildren.

After . . . well, after . . . my small sorrows didn't count. They weren't worth a mention. Only when my daughter was dangerously ill with pancreatitis did I become the focus—and thankfully only briefly. There were Diane's hysterical calls warning me what I must do to save Judy, what doctors to call, what to insist on. Diane viewed this as a second chance to save Karen. I understood. This time, Diane was not going to let death have its way.

CHAPTER TWENTY-ONE

I thought she told me everything. Now I know there were sensitive issues from the operator years that the New York crew, Jeffrey, and the cousins all knew, but that Diane didn't share with me. Did Mom and Dad know? Did Diane think they told me? Possibly by the time we spoke directly, the passage of time had resolved these issues. Or maybe Diane didn't want to pick at the scab of these troubles by bringing them up again. Maybe.

But now I wonder, were we as close as I thought? Do we ever really know someone?

While looking for details about the dressmaker's mirror accident for this book, I uncovered a family secret hidden for over one hundred years. I was astonished to find that my uncle Eugene's death certificate didn't mention an accident or injury of any kind. Rather, it stated that the four-year-old succumbed to congestive heart failure following a five-day stay at Carbondale (Pennsylvania) City Hospital. This was stunning because heart failure is a frequent complication of DCM, the disease that killed my sister and niece.

There was no accident. Eugene died because he inherited the deadly mutation. It is, however, unusual for an FLNC mutation to cause heart failure in such a young child. When I discussed this with Dr. McNally and Dr. Arad, they each thought it likely that Eugene suffered from two hits. The first, a weakening of his heart due to the FLNC mutation Karen had, and the second, either an additional mutation, or a virus that attacked his heart.

I am convinced that his parents—my grandparents Marion and David—told their sons the fictitious story about the mirror and that my father and his brother never questioned it. This pretense robbed my grandparents of the ability to talk about the true events and likely made the tragedy even harder for them to bear.

People in Carbondale well knew that heart failure caused little Eugene's death. I found articles about it in at least two local

My uncle Eugene Weiss's death certificate. He died from heart failure and not from injuries caused by an accident. PENNSYLVANIA VITAL RECORDS

newspapers. My grandparents could not remain in Carbondale if they wanted to promote a fiction of accidental death. Indeed, they soon left their Carbondale residence along with my grandfather's dream job with the railroad and started over in Scranton near my grandfather's brothers.

This secret had huge modern-day implications for my family. Why would my grandparents fabricate the heartbreaking tale of accidental death and blame it on my grandmother's negligence? I think they did it to protect their surviving son, my father, and their later issue, Uncle Cyrus, from being shunned by potential marriage partners. It was common for Jews to hide suspected hereditary defects. Possibly, they even knew of other relatives with similar ailments. Long before Watson and Crick discovered

CHAPTER TWENTY-ONE

DNA, my ancestors sensed and feared James Watson's message, "We used to think that our fate was in our stars, but now we know that, in large measure, our fate is in our genes."

Although Dad never knew the accident's grim secret, I think he set my course to become a research geneticist the day he told me about the tragedy. I recently completed a small research study that asked if the mutation in my family is a major cause of DCM among Ashkenazi Jews. Our results suggest that it is. After we published our data, which analyzed four Ashkenazi DCM patients,[4] we examined the FLNC gene in three more subjects. We only found seven Ashkenazim with DCM to study, because doctors don't diagnose most people with DCM until the end stage of the disease. Among these seven, three carried the same mutation as my family.

Doctors mistakenly told both my sister and one of these three subjects that a virus caused their DCM. To see if a viral story also misled other DCM patients, I asked a congestive heart failure Facebook group if anyone's doctor told them a virus caused their DCM. Within two hours of my post, I had seventeen positive responses. Doctors had referred only one of these seventeen for genetic testing. When I explained how to get genetic follow-up, the Facebook administrators removed my post.

I was frustrated when a study participant's doctor refused to order confirmatory clinical sequencing after we showed that the patient had the Ashkenazi FLNC mutation in the lab. I shared this experience with Dr. McNally and she responded, "I'm not surprised by the resistance you get from physicians about genetic testing. Most physicians have never had training in genetics and don't want to be burdened with one more thing. Despite this reluctance, we are seeing forward progress and the barriers are starting to come down.... So do keep pushing the message."

I used to blame my grandmother Marion's carelessness for Eugene's death. Now I see her as a vigilant mother, protecting her children from worrisome dangers, including her dressmaker's mirror. I am not surprised that she gave life to fears about the mirror when she needed a cover story to hide Eugene's true cause of death.

Marion's imaginary shattered mirror and the resulting broken glass shards embody her perceptive fears for her family's future. The premature deaths she envisioned came true with the passing of her mother at forty-seven; her own sudden death at fifty-nine; that of another son, my father, at sixty-six; the heart failure suffered by her granddaughter Diane in her fifties; and the sudden death of her great-granddaughter Karen at thirty-six, along with her unborn child. Despite Marion's premonition, there was nothing she could do.

When Marion, and later my father, died, it was unfortunate that the family didn't allow autopsies. Knowing that mother and son died of DCM rather than a heart attack could have made the genetic connection obvious by the time Diane developed the disease. This may have been enough to propel Karen's doctors to examine and treat her heart to prevent the sudden death. Also, of course, the family secret that little Eugene died of heart failure would have warned us. Although frowned upon by traditional Judaism, an autopsy can save a life which, according to Jewish law, trumps all other concerns.

Grandma Marion modeled for us how to deal with loss by choosing life in response to heartbreak. What she couldn't have known was that we would learn the cause of all the deaths, and that this would allow genetic testing and early treatment to end the family curse.

Advocacy for Genetic Testing

Genetic testing has undergone radical changes and advances in the past twenty years. Many physicians are still learning about these changes and factoring them into their medical practices. The reality is that mutations often cause disease in all ethnicities and that simple, relatively inexpensive genetic tests can save lives.

In the past, genetic counselors sometimes steered people away from testing because it cost thousands of dollars. Since prices have plummeted and benefits have multiplied, these patients as well as the public should now reconsider testing. Patients need only send a saliva sample.

Once sequencing finds the specific mutation responsible for a patient's disease, tests on other family members can determine whether they have the mutation. This allows those with the mutation to get preventive clinical care. Testing also frees those relatives without the mutation from worry.

Unfortunately, although mutations cause nearly 50 percent of all cardiomyopathy cases,[1] doctors currently refer only 1 percent of cardiomyopathy patients for testing.[2] As a result, most relatives never learn about the danger they face. Individuals with dilated cardiomyopathy (DCM) usually don't notice symptoms until the disease's late stage when they develop trouble breathing. Too often, the mutations lead to tragic, sudden deaths before people even become aware they have a problem.

We used to assume that if a patient with enlarged heart chambers (DCM) had a risk factor such as diabetes, lupus, or drug or alcohol misuse, that factor and not a mutation caused the heart problem. However, we now know that many such patients carry heart disease-causing mutations.[3]

The mutation that caused the deaths in my family is in the FLNC gene, which encodes a protein named filamin C, which heart muscles need to pump blood around the body. In my family a mutation changed the letter G into a C in an intron just before the exon starting at position 3791. This change destroyed the normal splicing of the FLNC mRNA and led to its degradation.[4] My laboratory research supports the conclusion that this mutation, called FLNC(3791-1G>C), first occurred many centuries ago in the Ashkenazi Jewish community. Today, one in eight hundred Ashkenazi Jews has it.[5]

It is important to understand that lots of other disease mutations in FLNC occur in *all* ethnicities. Indeed founder mutations also occur in other ethnicities.[6] In addition, mutations in over fifty other genes can cause the same heart ailment. Patients testing positive for one of these mutations should have their hearts tested periodically with a Holter monitor, cardiac MRI, and echocardiogram. If these detect a problem, their doctors could prescribe angiotensin-converting-enzyme (ACE) inhibitor or beta-blocker medications and, in advanced cases, implantation of a defibrillator or a heart transplant to prevent severe heart disease and thwart sudden death.

Popular DNA sequencing sites like 23andMe or Ancestry.com give interesting information about ethnicity and some genetic traits and can find new relatives. However, they are generally not suitable for detecting mutations that cause disease. This requires clinical genetic testing, which evaluates details of entire genes and determines if the changes found are likely harmful (pathogenic), benign, or of unknown significance. Companies repeat clinical sequencing many times before giving a result to the patient to be sure it is correct. A genetic counselor explains the results to patients. There are many DNA changes that are of unknown significance. In time these will be recharacterized as benign or pathogenic.

Some clinical tests focus on a panel of genes that cause a specific disease. As of 2024, self-pay options under $300, which include genetic counseling, are available for such panels. The advantage of using a panel is that you know precisely which of your genes the company sequenced. The disadvantage is that, since it is expensive to change panels, companies are slow to update them in response to new research. It took years of campaigning by Dr. McNally, myself, and perhaps others to get the FLNC gene that caused death in my family included in cardiomyopathy panels. Panel updating is critical because current genetic tests don't uncover the cause of cardiomyopathy in all families with a clear history of the disease. First-degree relatives of cardiomyopathy patients with a family history of the disease should continue enhanced clinical examinations and retest periodically until research uncovers their gene. In some families, the cause of disease may be complex, requiring more than one mutation to cause illness.[7] In the future, polygenic risk scores may enable relatives to gauge their risk even if the specific mutations responsible are unknown.[8]

An alternative approach to sequencing panels of genes is to instead sequence all of a patient's genes (exome sequencing) and sometimes even all the introns and everything in between the genes (whole genome sequencing). One Health Insurance Portability and Accountability Act of 1996 (HIPAA)–compliant company recently offered to do this sequencing for under $500 on sale, but they charge for the analysis of the sequence separately. Whichever method you use, you must remember that the interpretation of your sequences can change over time. This is especially true for negative or benign results.

Often, lacking a DNA sample from deceased relatives makes it difficult to complete the genetic story. To solve this problem, it is now easy and inexpensive to bank your DNA for future generations from a saliva sample. One such company gives you a tube with your DNA that is stable at room temperature for many decades.

Researchers estimate that patients with a pathogenic mutation in any of seven thousand genes are likely to develop corresponding diseases. Scientists know what some of these disease genes are and are looking for the rest of them. They are also working on treatments to reduce the mutations' harmful effects.

The ability of therapy to mitigate the effects of mutations in genes labeled "actionable" varies. As of June 2023, the American College of Medical Genetics and Genomics (ACMG) identified 81 disease genes as actionable,[9] while ClinGen reports that actions can be taken to mitigate the harmful effects of about 140 genes.[10] The ACMG recommends that if clinical sequencing incidentally uncovers mutations in genes on their actionable list, as part of secondary findings, doctors should inform the patient. Thanks in good measure to the story described in this book, including the work of Dr. McNally, Dr. Roth, Dr. Arad, Dr. Hershberger, and Dr. Gordon and the families they studied, the FLNC gene mutated in my family is now included in the ACMG actionable list. Mutations in actionable genes significantly increase the risk of various hereditary cancers, cardiovascular disease, and miscellaneous diseases. The number of genes included in these lists is growing.

The Healthy Nevada Project[11] offers free genetic screening to any Nevadan for the Centers for Disease Control and Prevention Tier 1 (CDCT1) genetic conditions.[12] The CDC chose the eleven genes on their list because of their high potential for actionability and impact on public health (e.g., Brca1 and Brca2). The Healthy Nevada Project found that current medical practice doesn't detect 90 percent of carriers for CDCT1 conditions, but widespread population genetic screening would find these people.[13]

Neither the ACMG nor ClinGen currently recommends proactive screening. That mutations found incidentally in actionable genes should be reported, but proactive population screening is not recommended seems inconsistent.[14] One distinction is the number of patients involved. ClinGen points out that we neither have enough physicians who are sufficiently trained for population-scale testing nor the follow-up required.[15]

It is essential that we solve these problems promptly. Real-world population-based clinical genetic screening found mutations that resulted in a change of care in one out of six people tested without symptoms or family history.[16] For example, one person in 250 in the general population has a mutation that can cause cardiomyopathy or sudden death before any symptoms appear.[17]

Several companies already offer proactive screens for $350 or less including a session with a genetic counselor. Some of these screen for almost all of the genes labeled actionable by the ACMG; others test far fewer genes. Companies provide the list of genes included in their test, which you can compare to the current ASGM or ClinGen actionable genes list. A healthcare professional needs to order these clinical genetic screens, but some companies have doctors who can do this for the patient proforma for under $20.

Even if the panel chosen is up to date at the time of the test, to get the best protection, patients should retest as the actionable lists grow. The approach of whole-exome sequencing rather than using targeted panels could solve this problem. Patients could learn if they had any mutations in the current actionable genes list and find out about more genes yearly without added sequencing. I expect companies to offer this in the next few years.

Finding a "pathogenic" mutation in a gene that could cause cancer, heart, or other diseases is not a definite prediction of disease. Rather, it means that the chance of disease is higher than for those without the mutation. The higher the penetrance and expressivity of the mutation, the higher the chance that the disease will actually develop. For some mutations, the lifetime chance of getting the disease may go way up, from 5 percent for persons without relevant mutations to 80 percent for those with a causative mutation. Mutations in the FLNC gene similar to the one found in my family cause a detectable change in the hearts of over 90 percent of people by age forty. This change could be heart arrhythmia or cardiomyopathy not yet noticed by the patient but sometimes followed by dire consequences if left untreated. For other mutations, the penetrance and expressivity is much lower but still of concern.

Doctors and patients need to interpret results of proactive screening without family history or symptoms with caution. The advice of a genetic counselor is essential; a mutation known to cause disease risk in one family might not be a risk in a different population. This is especially of concern for rare mutations.[18] A positive result calls for careful re-evaluation of family history, clinical assessment, and follow-up.

Acknowledgments

I sincerely thank my Rowman & Littlefield editor, Jacqueline Flynn, her assistant editors Joanna Wattenberg and Victoria Shi, as well as my wonderful literary agent Joan Parker for working closely with me and providing guidance and encouragement. Production editor Jessica Thwaite was essential to get this book across the finish line.

Special thanks are owed to Dr. Elizabeth McNally, Dr. Fredrick Roth, Dr. Michael Arad, and their teams for their groundbreaking work leading to the identification of the mutation in the previously unknown dilated cardiomyopathy heart disease gene, FLNC, that affected my family. I am also appreciative of their willingness to engage in discussions on science and medicine and to generously share unpublished data.

My heartfelt thanks go to Andrew Fried and Jeffrey Rothman for sharing their stories and writings, correcting errors, and improving the manuscript. I am also grateful to Cyrus Weiss for his nostalgic email about his childhood and valuable information regarding family history and Lisa Marie Castillo for her concern and the expert genetic counseling she provided to me and my family.

ACKNOWLEDGMENTS

I am indebted to Alice Wexler for reading and commenting on my book proposal. I greatly value the editing assistance and encouragement from Julie Sussman and Beth Carbone, as well as the helpful comments from my husband Alan Liebman, daughter Judith Liebman, son Michael Liebman, and cousins Mary Lynam, Barbara Zand, and Martin Weiss as well as help from Carmel Chiswick, Karen Van Slambrouck, Christine Paust, Lenore Lemon, Wendy Silk, Peggy Stinson, Alison Ripley Cubitt, Lisa Wright, and Sue Raymond. Thanks go to Joe Grzymski, Tom Petes, Min Li, and Yingying Tang for critiquing the sections on genetics.

The Jewish Literary Journal published short pieces of a few chapters of an earlier draft of this book as a creative nonfiction short story in their 100th online issue in October 2021.

APPENDIX 1

"Dear Family" Letter

Dr. Elizabeth McNally, MD, PhD, Director of Cardiovascular Genetics Clinic and Lisa Castillo, Genetics Counselor, both of the Center for Genetic Medicine at Northwestern Medical School, wrote the following official "Dear Family" letter for me to send to my family in February 2018. I removed the prices they listed in the original letter because they are now out of date and replaced personal account and telephone numbers with xxx for privacy.

A member of your family has been identified as carrying a genetic mutation that can predispose to heart conditions called hypertrophic cardiomyopathy (HCM) and dilated cardiomyopathy (DCM), and this form of cardiomyopathy can be linked to irregular heart rhythms that can be life-threatening. Hypertrophic cardiomyopathy means the wall of the left ventricle of the heart is thickened. Dilated cardiomyopathy means that the inside of the left ventricle is enlarged. The symptoms of DCM and HCM can be variable. Typically, symptoms range from no symptoms to symptoms that include shortness of breath, dizziness and/or fainting episodes. The shortness of breath may be progressive and worsen over time. Irregular heart rhythms can also occur and may happen without warning and could be life threatening. In this form of hypertrophic and dilated cardiomyopathy, irregular heart rhythms can occur in the absence of having a thickened left ventricle.

This is a genetic disorder. Hereditary information is passed down in our chromosomes to our children. Chromosomes are structures found in every cell of the body, which contain the hereditary material. There are 46 chromosomes in each cell of the body with 22 numbered pairs of chromosomes with the 23rd pair called the sex chromosomes (X or Y). An individual gets one chromosome of each pair from each parent. Chromosomes are made up of our genes. Genes contain the instructions that tell our bodies how to grow and develop. We carry two copies of every gene.

Currently there are more than 100 genes known to cause HCM and DCM. Both HCM and DCM can be inherited as an autosomal dominant disorder. Autosomal means that both males and females can be equally affected because the gene is not located on one of the sex chromosomes. Dominant inheritance means that one copy of the gene that is not working properly is enough to cause the condition. Because we only pass on one copy of each gene to children, there is a 50% chance that the copy with the mutation will be passed down and a 50% chance that the gene without the mutation will be passed down.

In your family, there is a gene that is not working properly because of a gene change (mutation) and this causes cardiomyopathy and irregular heart rhythms. The gene that has the mutation is the filamin C (FLNC) gene. The FLNC gene change in your family is c.3791-1G>C (IVS21-1 G>C). This translates to the position one before the 3791st position in the FLNC gene, there should be a letter G and instead there is the letter C. This change causes the protein made from this gene to stop prematurely and be an incomplete protein. Having a premature stop in FLNC can lead to disease such as cardiomyopathy and irregular heart rhythms. This change in FLNC has been reported as being disease causing as it has been found in other people with cardiomyopathy and irregular heart rhythms.

Truncating mutations in the FLNC gene have been associated with hypertrophic cardiomyopathy, dilated cardiomyopathy (DCM) and irregular heart rhythms that can be fatal. For people who have a known FLNC gene mutation, annual EKG, echocardiogram, Holter monitor or ZIO Patch and clinical cardiology evaluation is recommended. For those who have early signs of cardiomyopathy, there is potentially medical management that can slow the progression of disease. In addition, cardiac arrhythmias can be prevented by management with medications and sometimes devices.

Because you have received this letter, you may be at risk for having this FLNC gene mutation. If you would like to learn more about the gene, genetic testing such as the risks/benefits, and implications to testing, we would be happy to discuss this with you further.

The genetic testing can be accomplished with a blood test. The cost associated with this clinical test is approximately $xxx if using insurance or $xxx if paying out of pocket and is charged by the company that performs the test. This cost may be covered by insurance. The company that performs the genetic testing will typically check with your insurance company before testing to determine if your insurance is likely to cover this, and typically the testing company will call you prior to starting the test if the expected out of pocket is over $xxx with the ability to cancel the test or make it more affordable if over $xxx. We would be happy to coordinate the genetic testing, or if you do not live in the area, we can coordinate a mechanism to have the testing ordered.

Some people have questions about how genetic testing will affect their insurance. Under a federal law named GINA (Genetic

Information Non-discrimination Act), health insurance and employers cannot discriminate based on genetic testing information for those health insurance policies that are a large company plan (very small, private insurance is not covered under GINA). Life insurance is not protected under GINA or any other law, and therefore, having a genetic mutation can disqualify you from receiving life insurance. We recommend obtaining life insurance prior to undergoing genetic testing for individuals who do not have any symptoms or clinical diagnosis of DCM/HCM. For those who have a clinical diagnosis of DCM/HCM, genetic testing will not affect insurance if you already have a clinical diagnosis, and thus the diagnosis is already part of your medical record.

If/when you do proceed with genetic testing, please bring this letter in with you so that the physician/genetic counselor is aware of the exact mutation identified in your family and so that you are tested for the correct mutation. Your family member's genetic testing was ordered through two laboratories, one through a company called GeneDx and refer to accession number: XXXXX.

If you choose to not undergo gene testing, we would recommend that you have an EKG, echocardiogram, Holter monitor/ZIO Patch and evaluation with a cardiologist who is familiar with the risks associated with FLNC frameshifting gene mutations, and recognizes the arrhythmia risk prior to onset of cardiomyopathy. We realize that this can be a lot of information to take in and can be quite confusing. Please feel free to call us with any questions at XXX XXX-XXXX.

APPENDIX 2

Genetic Testing Sources

A good way to search for a genetic cause of a specific medical condition is to employ an appropriate gene panel. As of April 2024, GeneDx, Ambry Genetics, Invitae (in the process of being taken over by Labcorp), Blueprint Genetics, and Fulgent Genetics offer extensive panels specializing in cancer, cardiology, metabolic disorders, immunology, neurology, etc. Generally, each panel contain a few dozen to a few hundred genes. Self-pay prices for GeneDx, Ambry Genetics, and Invitae, are under $300, and this includes genetic counseling. Private insurance can often reduce the patient's cost to under $100. However, currently Medicare coverage for genetic testing is extremely limited. Blueprint and Fulgen Genetics genetic panels are more expensive and do not include genetic counseling for patients. However, they have a more extensive analysis of variants (mutations) of unknown significance and will sequence relatives' genes for free if that will help determine whether the variant is pathogenic or benign. Their genetic counselors will also talk with healthcare providers to help them choose appropriate tests and to explain results.

When companies uncover the gene variant that caused a patient's disease, they often offer reduced-cost or complimentary cascade testing for family members for that variant.

If panel tests don't find a pathogenic variant (mutation) known to cause the patient's diagnosis, doctors should plan to update the patient's testing yearly to take advantage of new knowledge. Over time, companies will reclassify variants they previously listed as having unknown significance (VUS) as being benign or pathogenic. Also, they will add new genes associated with the pathology to their panels.

If panel testing does not find the disease-causing mutation, doctors could try whole-exome sequencing. Sequencing all of the patient's approximately 20,000 genes along with his or her parents' genes (trio analysis) can, in combination with a family history, help identify which variants are associated with disease. The doctor supplies a clinical diagnosis and family history and the software combs through the sequencing results from all the genes to find the most relevant variants to report on. In addition, these tests will report on any positive secondary findings of pathogenic mutations in the ACMG eighty-one actionable genes. GeneDx, Ambry Genetics, Invitae, Fulgent Genetics, and PerkinElmer Genomics offer trio whole-exome sequencing for about $2,500 self-pay. Some of these tests come with offers of complimentary reanalysis in succeeding years. Some medical societies now recommend skipping the panel sequencing step and proceeding directly to whole-exome or genome sequencing to speed up diagnoses and treatment of neurodevelopmental disorders.

Invitae offers a proactive health screen panel for mutations that cause cancer, cardiovascular, and other medically active conditions for $299. Nebula Genomics offers whole-genomic sequencing directly to the consumer without involvement of a medical provider or medical insurance. They give you access to your sequence and the tools to explore it along with a lifetime of updated reports that include new discoveries for $500 to $1000. Dante has a similar offer.

While most of the above genetic tests include genetic counseling, some physicians prefer to have genetic counselors recommend tests prior to choosing a company. Genome Medical, Genome Ally, and Tandem Genetics provide such services. They meet with patients online to gather medical history and arrange for the appropriate tests, which are billed separately, coordinating insurance benefits.

Banking your DNA would allow descendants to analyze it to trace hereditary health conditions. Securigene currently provides this service.

RESOURCES

Ambry Genetics: www.ambrygen.com
Blueprint Genetics: www.blueprintgenetics.com
Dante labs: us.dantelabs.com
Fulgent Genetics: www.fulgentgenetics.com/products/disease/raredisease.html
GeneDx: www.genedx.com
Genome Ally: www.genomeally.com
Genome Medical: www.genomemedical.com
Invitae: www.invitae.com
Nebula Genomics: www.nebula.org
PerkinElmer Genomics: www.perkinelmergenomics.com
Securigene: www.securigene.com/dna-banking-preservation
Tandem Genetics: www.tandemgenetics.com/services/patient-services/

Notes

PREFACE

1. American Cancer Society, "Genetic Testing for Cancer Risk" www.cancer.org/cancer/risk-prevention/genetics/genetic-testing-for-cancer-risk.html (January 2, 2024); Kiran Musunuru et al., "Genetic Testing for Inherited Cardiovascular Diseases: A Scientific Statement from the American Heart Association," *Circulation: Genomic and Precision Medicine*, 13 (August 2020): 373–85.

2. Mauro Longoni et al., "Real-world utilization of guideline-directed genetic testing in inherited cardiovascular diseases," *Frontiers in Cardiovascular Medicine*, 10 (October 2023): 1272433–43.

3. Clinical Genome Resource, "Adult Actionability Workgroup protocol: Generation of summary reports and semi-quantitative metric," *ClinGen* 2020, www.clinicalgenome.org/site/assets/files/5074/adult_combined_evidence_curation_and_scoring_protocol_07292020.pdf (January 2, 2023).

4. ACMG Board of Directors, "The use of ACMG secondary findings recommendations for general population screening: A policy statement of the American College of Medical Genetics and Genomics (ACMG)," *Genetics in Medicine*, 21 (July 2019): 1467–68.

NOTES

THREE

1. Robert Guthrie, "Screening for phenylketonuria," *Triangle*, 9, no. 3 (1969): 104–109.
2. Health Resources and Services Administration, "Recommended Uniform Screening Panel," January 2023, www.hrsa.gov/advisory-committees/heritable-disorders/rusp
3. David Bick et al., "An online compendium of treatable genetic disorders," *American Journal of Medical Genetics C Seminar in Medical Genetics*, 187, no. 1 (March 2021): 48–54.
4. Ed Yong, "'We Gained Hope.' The Story of Lilly Grossman's Genome," *National Geographic*, March 11, 2013, www.nationalgeographic.com/science/article/we-gained-hope-the-story-of-lilly-grossmans-genome
5. Robert C. Green, "The BabySeq Project: Implementation of Whole Genome Sequencing as Screening in a Diverse Cohort of Healthy Infants," G2P Genomes to People (n.d.), www.genomes2people.org/research/babyseq/; Rady's Children's Institute for Genomic Medicine, "BeginNGS: Newborn Genomic Sequencing to end the diagnostic odyssey" (n.d.), radygenomics.org/begin-ngs-newborn-sequencing/
6. Zornitza Stark and Richard H. Scott, "Genomic newborn screening for rare diseases," *Nature Reviews Genetics*, 24 (November 2023): 755–66.

FOUR

1. Robert Waldinger and Marc Schulz, *The Good Life: Lessons from the World's Longest Scientific Study of Happiness* (New York: Simon & Schuster, 2023).

TEN

1. Nicholas J. Marini et al., "The prevalence of folate-remedial MTHFR enzyme variants in humans," *Proceedings of the Natl Academy*

of Sciences of the United States of America, 105, no. 23 (June 2008): 8055–60.

2. Daniela Concolino, Federica Deodato, and Rossella Parini, "Enzyme replacement therapy: efficacy and limitations," *Italian Journal of Pediatrics,* 44 (supplement 2), no. 120 (November 2018): 117–61.

3. Maria Francisca Coutinho, Juliana Inês Santos, Liliana Matos, and Sandra Alves, "Genetic substrate reduction therapy: A promising approach for lysosomal storage disorders," *Diseases,* 4, no. 4 (November 2016): 33–48.

4. Eleonora Riccio et al., "Switch from enzyme replacement therapy to oral chaperone migalastat for treating Fabry disease: Real-life data," *European Journal of Human Genetics,* 28 (July 2020): 1662–68.

5. Francesca Ferrua and Alessandro Aiuti, "Twenty-five years of gene therapy for ADA-SCID: From bubble babies to an approved drug," *Human Gene Therapy,* 28, no. 11 (November 2017): 972–81.

6. Ibid.

7. Tejal Aslesh and Toshifumi Yokota, "Restoring SMN expression: An overview of the therapeutic developments for the treatment of spinal muscular atrophy," *Cells,* 11, no. 3 (January 2022): 417–34.

8. Shimrit Oz, et al., "Reduction in Filamin C transcript is associated with arrhythmogenic cardiomyopathy in Ashkenazi Jews," *International Journal of Cardiology,* 317 (April 2020): 133–38.

9. Maria Cristina Valsecchi, "Rare diseases the next target for mRNA therapies," *Nature Italy,* May 9, 2021, www.nature.com/articles/d43978-021-00058-x.

10. Fyodor Urnov, "We can edit a person's DNA. So why don't we?" *New York Times,* December 11, 2022, 6(SR).

THIRTEEN

1. James F. Gusella et al., "A polymorphic DNA marker genetically linked to Huntington's disease," *Nature,* 306 (November 1983): 234–38.

NOTES

2. David Botstein, Raymond L. White, Mark Skolnick, and Ronald W. Davis, "Construction of a genetic linkage map in man using restriction fragment length polymorphisms," *American Journal of Human Genetics*, 32, no. 3 (May 1980): 314–31.

3. David T. Miller et al., "ACMG SF v3.2 list for reporting of secondary findings in clinical exome and genome sequencing: A policy statement of the American College of Medical Genetics and Genomics (ACMG)," *Genetics in Medicine*, 25, no. 8 (August 2023): 100866–72.

4. Clinical Genome Resource, "Actionablity knowledge repository," *ClinGene*, https://actionability.clinicalgenome.org/ac/Adult/ui/summ/assertion (January 4, 2024).

5. Iftikahar J. Kullo et al., "Polygenic scores in biomedical research," *National Review Genetics*, 23 (September 2022): 524–32.

6. ACMG Board of Directors, "The use of ACMG secondary findings recommendations for general population screening: A policy statement of the American College of Medical Genetics and Genomics (ACMG)," *Genetics in Medicine*, 21 (July 2019): 1467–68.

7. Robert Nussbaum, Eden Haverfield, Edward D. Esplin, and Swaroop Aradhya, "Response to 'The use of ACMG secondary findings recommendations for general population screening: A policy statement of the American College of Medical Genetics and Genomics (ACMG),'" *Genetics in Medicine*, 21, no. 12 (December 2019): 2836–37.

8. Clinical Genome Resource, "Adult Actionability Workgroup Protocol: Generation of Summary Reports and Semi-Quantitative Metric," *ClinGen* 2020, www.clinicalgenome.org/site/assets/files/5074/adult_combined_evidence_curation_and_scoring_protocol_07292020.pdf (January 2, 2024).

9. Anne Andermann, Ingeborg Blancquaert, Sylvie Beauchamp, and Veronique Dery, "Revisiting Wilson and Jungner in the genomic age: A review of screening criteria over the past 40 years," *Bulletin of the World Health Organization*, 86, no. 4 (April 2008): 317–19.

10. National Human Genome Research Insitute, "Policy Issues in Genomics," www.genome.gov/about-genomics/policy-issues (January 6, 2022).

NOTES

NINETEEN

1. Andrew Fried, "The life I knew and its shattering," *Fry Guy's Thinkerings,* December 1, 2008, fryguysthinkings.blogspot.com/2008/12/life-i-knew-and-its-shattering.html (January 4, 2024).

2. Wystan Hugh Auden, "Twelve Songs – IX," *W. H. Auden: Collected Poems* (Penguin Random House LLC, 2007).

TWENTY

1. Jessica R. Golbus et al., "Targeted analysis of whole genome sequence data to diagnose genetic cardiomyopathy," *Circulation: Cardiovascular Genetics,* 7, no. 6 (December 2014): 751–59.

2. Anna Abramowicz and Monika Gos, "Splicing mutations in human genetic disorders: Examples, detection, and confirmation," *Journal of Applied Genetics,* 59, no. 3 (August 2018): 253–68.

3. Shimrit Oz et al., "Reduction in Filamin C transcript is associated with arrhythmogenic cardiomyopathy in Ashkenazi Jews," *International Journal of Cardiology,* 317 (April 2020): 133–38.

4. Rahul C. Deo et al., "Prioritizing causal disease genes using unbiased genomic features," *Genome Biology,* 15, no. 12 (December 2014): 534–38.

5. ClinVar Miner. https://clinvarminer.genetics.utah.edu/. Accessed December 12, 2023.

6. Martín F. Ortiz-Genga et al., "Truncating FLNC mutations are associated with high-risk dilated and arrhythmogenic cardiomyopathies," *Journal of the American College of Cardiology,* 68, no. 22 (December 2016): 2440–51.

7. Andre Goffeau et al., "Life with 6000 genes," *Science,* 247, no. 5287 (October 1996): 546–67.

8. Sergey Nurk et al., "The complete sequence of a human genome," *Science,* 376, no. 6588 (April 2022): 44–53.

9. Shai Carmi et al. "Sequencing an Ashkenazi reference panel supports population-targeted personal genomics and illuminates

Jewish and European origins," *Nature Communications*, 5. no. 4835 (2014). https://doi.org/10.1038/ncomms5835

10. Andrea Ganna et al., "Quantifying the impact of rare and ultra-rare coding variation across the phenotypic spectrum," *American Journal of Human Genetics*, 102, no. 6 (June 2018): 1204–211; Manuel A. Rivas et al., "Insights into the genetic epidemiology of Crohn's and rare diseases in the Ashkenazi Jewish population," *PLoS Genetics*, 14, no. 5 (May 2018): e1007329, doi: 10.1371/journal.pgen.1007329.

11. Abhinav Jain, Disha Sharma, Anjali Bajaj, Vishu Gupta, and Vinod Scaria, "Founder variants and population genomes—Toward precision medicine," *Advances in Genetics*, 107, no. 4 (2021): 121–52.

12. Martín F. Ortiz-Genga et al., "Truncating FLNC mutations are associated with high-risk dilated and arrhythmogenic cardiomyopathies," *Journal of the American College of Cardiology*, 68, no. 22 (December 2016): 2440–51.

TWENTY-ONE

1. Jessica R. Golbus et al., "Targeted analysis of whole genome sequence data to diagnose genetic cardiomyopathy," *Circulation: Cardiovascular Genetics*, 7, no. 6 (September 2014): 751–59; Martin F. Ortiz-Genga et al., "Truncating FLNC mutations are associated with high-risk dilated and arrhythmogenic cardiomyopathies," *Journal of the American College of Cardiology*, 68, no. 22 (December 2016): 2440–51.

2. Martin F. Ortiz-Genga et al., "Truncating FLNC mutations are associated with high-risk dilated and arrhythmogenic cardiomyopathies," *Journal of the American College of Cardiology*, 68, no. 22 (December 2016): 2440–51.

3. Excerpts of the March 2018 letter I wrote to my nephew about the family mutation: (a) For these mutations the first symptoms are often not dilated cardiomyopathy (DCM) or heart failure nor do they show up on an echocardiogram (although often these symptoms do occur). Rather, for this type of FLNC truncating mutation, the first

symptom is often intramyocardial fibrosis and ventricular arrhythmias that can cause sudden death (SD). Because of this, it is recommended that Holter monitoring and MRI (for the evaluation of intramyocardial fibrosis) be included in the clinical evaluation of mutation carriers. The very helpful doctor wrote to me: "After our investigation, we have learned that truncating mutations in FLNC can produce intramyocardial fibrosis and ventricular arrhythmias with an unremarkable echocardiogram. So, we always recommend to screen 'asymptomatic' carriers with periodical cardiac-MRI (could detect fibrosis through late gadolinium enhancement) and Holter ECG." (b) Carriers with symptoms (e.g., irregular heartbeat) found in Holter test are suggested to get a defibrillator implant. (c) In one of the families with our exact mutation, a grandmother with the mutation lived to seventy-nine with no heart symptoms (she died of Alzheimer's), but her son with the mutation, who in this case probably had symptoms of DCM, died of SD at thirty-two. Her daughter with the mutation had a symptom of sustained ventricular tachycardia (SVT) starting at age fifty. This daughter had two sons who both inherited the mutation. One died suddenly at age seventeen with no prior symptoms, when playing handball. The other son at twenty-four has no symptoms. (d) The other two families with our exact mutation are also of Ashkenazi Jewish descent. This mutation has not been found in any other ethnicities. Their two families have no other closer relationships than being Ashkenazi. I sent them the history of our family, and no relationships were found with either of their two families. It is thus likely that this is a new "Ashkenazi-Jewish" gene. (e) To get tested for the mutation, you simply need any doctor to place an order (my doctor has offered to do it; see attached). If you do it thru GeneDx, they already have my niece's DNA to use as a control. Without going through insurance this would cost $245, and with insurance it may cost much less or nothing (although they will charge the insurance more). All you do is spit in a tube. If you go with GeneDx, have your doctor ask for a carrier test for our specific mutation (or better yet have them call my doctor as instructed in letter). GeneDx should do for you exactly what they did for me, Susan Liebman. Use the numbers and info in the Family Letter attached.

4. Susan W. Liebman, Haley Palaganas, and Haviva Kobany, "A founder mutation in FLNC is likely a major cause of idiopathic dilated cardiomyopathy in Ashkenazi Jews," *International Journal of Cardiology*, 323 (January 2021): 124.

ADVOCACY FOR GENETIC TESTING

1. Jeffrey A. Towbin, "Inherited cardiomyopathies," *Circulation Journal*, 78, no. 10 (September 2014).
2. Mauro Longoni et al., "Real-world utilization of guideline-directed genetic testing in inherited cardiovascular diseases," *Frontiers in Cardiovascular Medicine*, 10 (October 2023): 1272433-43.
3. Job A. J. Verdonschot et al., "Implications of genetic testing in dilated cardiomyopathy," *Circulation. Genomic Precision Medicine*, 13, no. 5 (September 2020): 476-87; Takanobu Yamada and Seitaro Nomura, "Recent findings related to cardiomyopathy and genetics," *International Journal of Molecular Sciences*, 22, no. 22 (November 2021): 12522-34.
4. Shimrit Oz et al., "Reduction in Filamin C transcript is associated with arrhythmogenic cardiomyopathy in Ashkenazi Jews," *International Journal of Cardiology*, 317 (April 2020): 133-38.
5. Susan W. Liebman, Haley Palaganas, and Haviva Kobany, "A founder mutation in FLNC is likely a major cause of idiopathic dilated cardiomyopathy in Ashkenazi Jews," *International Journal of Cardiology*, 323 (January 2021): 124.
6. Abhinav Jain, Disha Sharma, Anjali Bajaj, Vishu Gupta, and Vinod Scaria, "Founder variants and population genomes—Toward precision medicine," *Advances in Genetics*, 107, no. 4 (2021): 121-52.
7. Elizabeth Jordan and Ray E. Hershberger, "Considering complexity in the genetic evaluation of dilated cardiomyopathy," *Heart*, 107, no. 2 (January 2021): 106-112.
8. Iftikahar J. Kullo et al., "Polygenic scores in biomedical research," *Nature Reviews. Genetics*, 23 (September 2022): 524-32.
9. David T. Miller et al., "ACMG SF v3.2 list for reporting of secondary findings in clinical exome and genome sequencing: A

policy statement of the American College of Medical Genetics and Genomics (ACMG)," *Genetics in Medicine*, 25, no. 8 (August 2023): 100866–72; David T. Miller et al., "ACMG SF v3.0 list for reporting of secondary findings in clinical exome and genome sequencing: A policy statement of the American College of Medical Genetics and Genomics (ACMG)," *Genetics in Medicine*, 23 (May 2021): 1381–90.

10. Clinical Genome Resource, "Actionablity knowledge repository," *ClinGene*, actionability.clinicalgenome.org/ac/Adult/ui/summ/assertion (accessed January 4, 2024).

11. Healthy Nevada Project, healthynv.org (accessed January 5, 2024).

12. William D. Dotson et al., "Prioritizing genomic applications for action by level of evidence: A horizon-scanning method," *Clinical Pharmacology & Therapeutics*, 95, no. 4 (April 2014): 394–402.

13. Joseph J. Grzymski et al., "Population genetic screening efficiently identifies carriers of autosomal dominant diseases," *Nature Medicine*, 26, no. 8 (August 2020): 1235–39.

14. Robert Nussbaum, Eden Haverfield, Edward D. Esplin, and Swaroop Aradhya, "Response to 'The use of ACMG secondary findings recommendations for general population screening: A policy statement of the American College of Medical Genetics and Genomics (ACMG),'" *Genetics in Medicine*, 21, no. 12 (December 2019): 2836–37.

15. Clinical Genome Resource, "Adult Actionability Workgroup Protocol: Generation of Summary Reports and Semi-Quantitative Metric," *ClinGen* 2020, www.clinicalgenome.org/site/assets/files/5074/adult_combined_evidence_curation_and_scoring_protocol_07292020.pdf (accessed January 2, 2024).

16. Eden V. Haverfield et al., "Physician-directed genetic screening to evaluate personal risk for medically actionable disorders: A large multi-center cohort study," *BMC Medicine*, 19, no. 1 (November 2021): PMID: 34404389.

17. Heinz-Peter Schultheiss et al., "Dilated cardiomyopathy," *Nature Reviews. Disease Primers*, 5, no. 1 (May 2019): 32–51.

18. Elizabeth Jordan and Ray E. Hershberger, "Considering complexity in the genetic evaluation of dilated cardiomyopathy," *Heart*, 107, no. 2 (January 2021): 106–112.

Glossary of Jewish Words

Afikomen—A piece of matzah hidden during the Passover Seder and later found by children. It is the last thing eaten at the seder meal. It comes from a Greek word meaning *dessert*.

Ashkenazi—Descendants of less than five hundred medieval Jews in Central and Eastern Europe whose genetic origins were equally split between the Levant and Europe.

Aufruf—A ceremony in which an engaged couple is called up to the Torah before their wedding.

Bar mitzvah—Ceremony for a Jewish boy turning thirteen, when he has all the rights and obligations of a Jewish adult.

Benching licht—Lighting Shabbat candles.

Borscht—A traditional Eastern European soup made from beets.

Bracha—A blessing recited before or after performing a mitzvah or eating.

Bris—A ritual circumcision performed on male infants.

Challah—A braided bread traditionally eaten on Shabbat and holidays.

Diyanu—A Passover song expressing gratitude for the blessings in life.

Haggadah—A Jewish book that guides participants through the rituals and prayers at the Passover seder.

Kaddish—A prayer recited by mourners in memory of the deceased.
Kasha—Buckwheat groats, often used in Jewish cuisine.
Kedush—Holiness or sanctification.
Kippah—A skullcap, sometimes called a *yarmulke,* worn by Jewish men as a sign of reverence to God.
Kohanim—The descendants of Aaron who have special priestly duties in Judaism.
Kosher—Foods that comply with strict Jewish dietary laws and that are prepared, combined, and consumed according to these laws are kosher.
Kreplach—Small dumplings typically filled with meat and served in soup.
Leviim—The Levites, descendants of a tribe of ancient Israelites with religious responsibilities.
Manischewitz—A brand of kosher wine often associated with Jewish celebrations.
Marit ayin—The Jewish principle to avoid even the appearance of wrongdoing.
Matzah—Unleavened bread eaten during the Passover festival when Jews are forbidden to eat leaven.
Minyan—A quorum of ten Jewish adults required for certain communal prayers.
Mitzvah (plural Mitzvot)—A commandment from God.
Mohel—A person trained in the ritual circumcision of male infants.
Orthodox Judaism—Branch of Judaism that more closely follows ancient rules than do the Conservative and Reform Jewish branches.
Passover—A Jewish holiday that celebrates the freedom of Israelites from Egyptian slavery.
Pogrom—A violent mob attack on religious minority.
Schlep—To drag or carry something with effort.

Schmatte—A rag or scrap of cloth, often used to refer to old or worn-out clothing.

Seder—Ritual service that follows a specific order including a festival meal that occurs on the first night or two of the Passover holiday.

Shabbat—The Jewish day of rest and observance, beginning on Friday at sundown.

Sheitel—A wig worn by Orthodox Jewish women to hide their hair from men other than their husbands as prescribed by Jewish law.

Shiva—The seven-day mourning period observed after the death of a close relative.

Simcha—A happy event, often a joyous celebration.

Shtetl—The generic name for largely Jewish small towns that existed in Eastern Europe.

Torah—Scroll of the Hebrew Bible.

Tzaddik—A righteous and pious person; the spiritual leader of a Hasidic community.

Yahrzeit—Jewish commemoration of the anniversary of a loved one's death.

Yisraelim—Jewish people, the collective of Israelites.

Glossary of Scientific and Medical Terms

Amino acid—Building block of protein; twenty different amino acids are used to make protein.

Bilirubin—A yellow pigment produced during the breakdown of red blood cells.

Centromere—A specific region of a chromosome that plays a role in cell division and helps in the proper segregation of the chromosome.

Chromosome—A structure found in the nucleus that contains one long piece of DNA containing hundreds to thousands of genes.

Codon—A sequence of three of the four RNA "letters" in messenger RNA (mRNA). The sixty-one codons that code for twenty specific amino acids during protein synthesis are *sense* codons. The three codons used to signal termination of the protein are *nonsense* or *stop* codons.

DCM—Dilated Cardiomyopathy, a condition characterized by the enlargement and weakened function of the heart muscle.

DNA—Deoxyribonucleic Acid. It is made of long chains of "letters" abbreviated A, T, G, and C; the order of the letters encodes inherited traits.

Ejection fraction—The fraction of blood the heart pumps out of the left ventricle with every heartbeat.

Enzyme—A protein that speeds up a chemical reaction in the cell.

Eukaryote—An organism whose cells have a nucleus, including plants, animals, fungi (including yeast), and protists.

Exome—All of the exons; this is about 1 percent of the human genome and excludes sequences that do not code for protein. The excluded sequences are found in between genes and in the "introns" within genes that separate different exons (protein coding regions) within a single gene.

Exons—DNA sequences that directly code for proteins.

Expressivity—The degree to which a trait is expressed in people with a particular mutation.

FLNC—The FLNC gene encodes the filamin C protein, which functions in skeletal and cardiac muscles.

Gene—A region of DNA that controls and encodes synthesis of a specific protein or other basic component of the cell such as ribosomal or transfer RNA.

Genome—All (100 percent) of the DNA in an individual including the DNA that does not code for protein that is found in between genes and in the "introns" within genes that separate different exons (protein coding regions) within a single gene.

Holter test—A diagnostic test that records the electrical activity of the heart over a period of time, typically twenty-four to forty-eight hours, to detect abnormal heart rhythms.

Introns—Sequences that separate different exons (protein coding regions) within a single gene.

Microtubule—Cylindrical structure made of the protein tubulin that attaches to a centromere and pulls duplicated chromosomes to opposite sides of the cell during cell division.

Mutation—A permanent alteration in the DNA sequence, which can lead to changes in traits or diseases.

Penetrance—The fraction of people with a mutation that express a trait associated with the mutation.

Prion—An infectious shape of a protein that converts other molecules of that protein into the prion shape.

Promoter—A DNA sequence that plays a crucial role in initiating transcription of genes into messenger RNA (mRNA) by binding transcription factors and RNA polymerase.

Protein—A chain of hundreds or thousands of amino acids. The sequence of the amino acids determines the protein's structure and function.

[*PSI+*]—A prion form of the yeast SUP35 protein that causes it to lose the ability to find the ends of genes, thereby resulting in observable changes in yeast cells.

QT syndrome—A group of disorders characterized by an abnormality in the heart's electrical activity, leading to an increased risk of dangerous heart rhythms.

Ribosome—A cellular structure responsible for protein synthesis that translates messenger RNA(mRNA) into proteins.

RNA—Ribonucleic Acid. It is made of long chains of "letters" abbreviated A, U, G, and C transcribed from a section of DNA. There are different types of RNA, including: messenger RNA (mRNA), which codes for protein; ribosomal RNA (rRNA), which is a structural component of the ribosome; and transfer RNA (tRNA), which is involved in synthesizing protein.

Spherocytosis—A hereditary disorder characterized by the presence of spherically-shaped red blood cells, leading to anemia and jaundice.

Variant—Different versions of the same gene each containing changes (mutations) that may or may not have an effect on the function of the encoded protein.

Virion—A complete virus particle consisting of nucleic acid (DNA or RNA) surrounded by a protein coat (capsid).

Bibliography

Abramowicz, Anna, and Monika Gos. "Splicing mutations in human genetic disorders: examples, detection, and confirmation." *Journal of Applied Genetics*, 59, no. 3 (August 2018): 253-68.

American Cancer Society. "Genetic testing for cancer risk." www.cancer.org/cancer/risk-prevention/genetics/genetic-testing-for-cancer-risk.html (accessed January 2, 2024).

Andermann, Anne, Ingeborg Blancquaert, Sylvie Beauchamp, and Veronique Dery. "Revisiting Wilson and Jungner in the genomic age: A review of screening criteria over the past 40 years." *Bulletin of the World Health Organization*, 86, no. 4 (April 2008): 317-19.

Aslesh, Tejal, and Toshifumi Yokota. "Restoring SMN expression: An overview of the therapeutic developments for the treatment of spinal muscular atrophy." *Cells*, 11, no. 3 (January 2022): 417-34.

Auden, Wystan Hugh. "Twelve Songs – IX," *W. H. Auden: Collected Poems*. Penguin Random House LLC, 2007.

ACMG Board of Directors. "The use of ACMG secondary findings recommendations for general population screening: A policy statement of the American College of Medical

Genetics and Genomics (ACMG)." *Genetics in Medicine*, 21 (July 2019): 1467-68.

Bick, David, Sarah L. Bick, David P. Dimmock, Tom A. Fowler, Mark J. Caulfield, and Richard H. Scottet. "An online compendium of treatable genetic disorders." *American Journal of Medical Genetics, C Seminar Medical Genetics*, 187, no. 1 (March 2021): 48-54.

Botstein, David, Raymond L. White, Mark Skolnick, and Ronald W. Davis. "Construction of a genetic linkage map in man using restriction fragment length polymorphisms." *American Journal of Human Genetics*, 32, no. 3 (May 1980): 314-31.

Carmi, Shai, et al. "Sequencing an Ashkenazi reference panel supports population-targeted personal genomics and illuminates Jewish and European origins." *Nature Communications*, 5, no. 4835 (2014). https://doi.org/10.1038/ncomms5835.

Clinical Genome Resource. "Adult Actionability Workgroup protocol: Generation of summary reports and semi-quantitative metric scoring." *ClinGen* 2020, www.clinicalgenome.org/site/assets/files/5074/adult_combined_evidence_curation_and_scoring_protocol_07292020.pdf (accessed January 2, 2024).

Clinical Genome Resource. "Actionablity knowledge repository," *ClinGene*, actionability.clinicalgenome.org/ac/Adult/ui/summ/assertion (accessed January 4, 2024).

ClinVar Miner. https://clinvarminer.genetics.utah.edu/ (accessed December 12, 2023).

Concolino, Daniela, Federica Deodato, and Rossella Parini. "Enzyme replacement therapy: Efficacy and limitations." *Italian Journal of Pediatrics*, 44(supplement 2), no. 120 (November 2018): 117-61.

Coutinho, Maria Francisca, Juliana Inês Santos, Liliana Matos, and Sandra Alves. "Genetic substrate reduction therapy: A promising approach for lysosomal storage disorders." *Diseases*, 4, no. 4 (November 2016): 33-48.

Deo, Rahul C., Gabriel Musso, Murat Tasan, and Paul Tang. "Prioritizing causal disease genes using unbiased genomic features." *Genome Biology,* 15, no. 12 (December 2014): 534–38.

Dotson, William D., Michale P. Douglas, Katherine Kolor, Ann-Charlotte Stewart, Scott Bowen, Marta Gwinn, Anja Wulf et al. "Prioritizing genomic applications for action by level of evidence: a horizon-scanning method." *Clinical Pharmacology & Therapeutics,* 95, no. 4 (April 2014): 394–402.

Ferrua, Francesca, and Alessandro Aiuti. "Twenty-five years of gene therapy for ADA-SCID: From bubble babies to an approved drug." *Humam Gene Therapy,* 28, no.11 (November 2017): 972–81.

Fried, Andrew. "The life I knew and its shattering." *Fry Guy's Thinkerings,* December 1, 2008, fryguysthinkings.blogspot.com/2008/12/life-i-knew-and-its-shattering.html.

Ganna, Andrea, F. Kyle Satterstrom, Seyedeh M. Zekavat, Indraniel Das, Mitja I. Kurki, Claire Churchhouse, Jessica Alfoldi et al. "Quantifying the impact of rare and ultra-rare coding variation across the phenotypic spectrum." *American Journal of Human Genetics,* 102, no. 6 (June 2018): 1204–211.

Goffeau, Andre, Bart G. Barrell, Howard Bussey, Ronald W. Davis, Bernard Dujon, Horst Feldmann, Francis Galibert et al. "Life with 6000 genes." *Science,* 247, no. 5287 (October 1996): 546–67.

Golbus, Jessica R., Megan J. Puckelwartz, Lisa Dellefave-Castillo, John P. Fahrenbach, Viswateja Nelakuditi, Lorenzo L. Pesce, Peter Pytel, and Elizabeth M. McNally. "Targeted analysis of whole genome sequence data to diagnose genetic cardiomyopathy." *Circulation, Cardiovascular Genetics,* 7, no. 6 (December 2014): 751–59.

Green, Robert C. "The BabySeq Project: Implementation of whole genome sequencing as screening in a diverse cohort of healthy infants." G2P Genomes to People. www.genomes2people.org/research/babyseq (accessed January 1, 2024).

Grzymski, Joseph, Gai Elhanan, Joel A. Morales-Rosado, E. Smith, Karen Schlauch, Robert Read, C. Rowan et al. "Population genetic screening efficiently identifies carriers of autosomal dominant diseases." *Nature Medicine,* 26, no. 8 (August 2020): 1235–39.

Gusella, James F., Nancy S. Wexler, P. Michael Conneally, Susan L. Naylor, Mary Anne Anderson, Rudolph E. Tanzi, Paul C. Watkins et al. "A polymorphic DNA marker genetically linked to Huntington's disease." *Nature,* 306 (November 1983): 234–38.

Guthrie, Robert. "Screening for phenylketonuria." *Triangle,* 9, no. 3 (1969): 104–109.

Haverfield, Eden V., Edward D. Esplin, Sienna J. Aguilar, Kathryn E. Hatchell, Kelly E. Ormond, Andrea Hanson-Kahn, Paldeep S. Atwal et al. "Physician-directed genetic screening to evaluate personal risk for medically actionable disorders: A large multi-center cohort study." *BMC Medicine,* 19, no. 1 (November 2021): PMID: 34404389.

Health Resources and Services Administration. "Recommended Uniform Screening Panel," January 2023. www.hrsa.gov/advisory-committees/heritable-disorders/rusp.

Healthy Nevada Project. healthynv.org (accessed January 5, 2024).

Jain, Abhinav, Disha Sharma, Anjali Bajaj, Vishu Gupta, and Vinod Scaria. "Founder variants and population genomes—Toward precision medicine." *Advances in Genetics,* 107, no. 4 (2021): 121–52.

Jordan, Elizabeth, and Ray E. Hershberger. "Considering complexity in the genetic evaluation of dilated cardiomyopathy." *Heart,* 107, no. 2 (January 2021): 106–112.

Kullo, Iftikhar J., Cathryn M. Lewis, Michael Inouye, Alicia R. Martin, Samuli Ripatti, Nilanjan Chatterjee et al. "Polygenic scores in biomedical research." *Nature Reviews, Genetics,* 23 (September 2022): 524–32.

Liebman, Susan W., Haley Palaganas, and Haviva Kobany. "A founder mutation in FLNC is likely a major cause of idiopathic dilated cardiomyopathy in Ashkenazi Jews." *International Journal of Cardiology*, 323 (January 2021): 124.

Longoni, Mauro, Kanchan Bhasin, Andrew Ward, Donghyun Lee, McKenna Nisson, Sucheta Bhatt, Fatima Rodriguez, and Rajesh Dash. "Real-world utilization of guideline-directed genetic testing in inherited cardiovascular diseases." *Frontiers in Cardiovascular Medicine*, 10 (October 2023): 1272433-43.

Marini, Nicholas J., Jennifer Gin, Janet Ziegle, Kathryn Hunkapiller Keho, David Ginzinger, Dennis A. Gilbert, and Jasper Rine. "The prevalence of folate-remedial MTHFR enzyme variants in humans." *Proceedings of the National Academy of Sciences of the United States of America*, 105, no. 23 (June 2008): 8055-60.

Miller, David T., Kristy Lee, Adam S Gordon, Laura M. Amendola, Kathy Adelman, Sherri J. Bale, Wendy K. Chung et al. "ACMG SF v3.2 list for reporting of secondary findings in clinical exome and genome sequencing: A policy statement of the American College of Medical Genetics and Genomics (ACMG)." *Genetics in Medicine*, 25, no. 8 (August 2023): 100866-72.

Miller, David T., Kristy Lee, Wendy K. Chung, Adam S. Gordon, Gail E. Herman, Teri E. Klein, Douglas R. Stewart et al., "ACMG SF v3.0 list for reporting of secondary findings in clinical exome and genome sequencing: A policy statement of the American College of Medical Genetics and Genomics (ACMG)." *Genetics in Medicine*, 23 (May 2021): 1381-90.

Musunuru, Kiran, Ray E. Hershberger, Sharlene M. Day, N. Jennifer Klinedinst, Andrew P. Landstrom, Victoria N Parikh, Siddharth Prakash, Christopher Semsarian, and Amy C. Sturm. "Genetic testing for inherited cardiovascular diseases: A scientific statement from the American Heart Association." *Circulation, Genomic and Precision Medicine*, 13 (August 2020): 373-85.

National Human Genome Research Insitute. "Policy issues in genomics." www.genome.gov/about-genomics/policy-issues (accessed January 6, 2022).

Nurk, Sergey, Sergey Koren, Arang Rhie, Mikko Rautiainen, Andrey V. Bzikadze, Alla Mikheenko, Mitchell R. Vollger et al. "The complete sequence of a human genome." *Science*, 376, no. 6588 (April 2022): 44–53.

Nussbaum, Robert, Eden Haverfield, Edward D. Esplin, and Swaroop Aradhya. "Response to 'The use of ACMG secondary findings recommendations for general population screening: a policy statement of the American College of Medical Genetics and Genomics (ACMG).'" *Genetics in Medicine*, 21, no. 12 (December 2019): 2836–37.

Ortiz-Genga, Martin F., Sofía Cuenca, Matteo Dal Ferro, Esther Zorio, Ricardo Salgado-Aranda, Vicente Climent, Laura Padrón-Barthe et al. "Truncating FLNC mutations are associated with high-risk dilated and arrhythmogenic cardiomyopathies." *Journal of the American College of Cardiology*, 68, no. 22 (December 2016): 2440–51.

Oz, Shimrit, Hagith Yonath, Leonid Visochyk, Efrat Ofek, Natalie Landa, Haike Reznik-Wolf, Martin Ortiz-Genga et al. "Reduction in Filamin C transcript is associated with arrhythmogenic cardiomyopathy in Ashkenazi Jews." *International Journal of Cardiology*, 317 (April 2020): 133–38.

Rady's Children's Institute for Genomic Medicine. "BeginNGS: Newborn Genomic Sequencing to end the diagnostic odyssey." radygenomics.org/begin-ngs-newborn-sequencing/ (accessed January 1, 2024).

Riccio, Eleonora, Mario Zanfardino, Lucia Ferreri, Ciro Santoro, Sirio Cocozza, Ivana Capuano, Massimo Imbriaco, Sandro Feriozzi, Antonio Pisani, and AFFIINITY Group. "Switch from enzyme replacement therapy to oral chaperone migalastat for treating Fabry disease: Real-life data." *European Journal of Human Genetics*, 28 (July 2020): 1662–68.

Rivas, Manuel A., Brandon E. Avila, Jukka Koskela, Hailiang Huang, Christine Stevens, Matti Pirinen, Talin Haritunians et al. "Insights into the genetic epidemiology of Crohn's and rare diseases in the Ashkenazi Jewish population." *PLoS Genetics*, 14, no. 5 (May 2018): e1007329, doi: 10.1371/journal.pgen.1007329.

Schultheiss, Heinz-Peter, DeLisa Fairweather, Alida L. P. Caforio, Felicitas Escher, Ray E. Hershberger, Steven E. Lipshultz, Peter P. Liu et al. "Dilated cardiomyopathy." *Nature Reviews, Disease Primers*, 5, no. 1 (May 2019): 32–51.

Stark, Zornitza, and Richard H. Scott, "Genomic newborn screening for rare diseases." *Nature Reviews, Genetics*, 24 (November 2023): 755–66.

Towbin, Jeffrey A. "Inherited cardiomyopathies." *Circulation Journal*, 78, no. 10 (September 2014).

Urnov, Fyodor. "We can edit a person's DNA. So why don't we?" *New York Times*, Dec. 11, 2022, 6(SR).

Valsecchi, Maria Cristina. "Rare diseases the next target for mRNA therapies," *Nature Italy* 2021, www.nature.com/articles/d43978-021-00058-x.

Verdonschot, Job A. J., Mark R. Hazebroek, Ingrid P. C. Krapels, Michiel T. H. M. Henkens, Anne Raafs, Ping Wang, Jort J. Merken et al. "Implications of genetic testing in dilated cardiomyopathy." *Circulation, Genomic Precision in Medicine*, 13, no. 5 (September 2020): 476–87.

Waldinger, Robert, and Marc Schulz. *The Good Life: Lessons from the World's Longest Scientific Study of Happiness* (New York: Simon & Schuster, 2023).

Yamada, Takanobu, and Seitaro Nomura. "Recent findings related to cardiomyopathy and genetics." *International Journal of Molecular Sciences*, 22, no. 22 (November 2021): 12522-34.

Yong, Ed. "'We Gained Hope.' The story of Lilly Grossman's genome." *National Geographic* 2013, www.nationalgeographic.com/science/article/we-gained-hope-the-story-of-lilly-grossmans-genome.

About the Author

Susan W. Liebman, PhD, is a geneticist and the member of a family shattered by a deadly mutation. She was among MIT's early female undergraduates, later earning her PhD at the University of Rochester. She taught genetics at the University of Illinois at Chicago for thirty-four years while leading a continuously grant-funded research group. During her tenure there, she was the first woman to be awarded the title of Distinguished University Professor. Currently, Dr. Liebman is a research professor at the University of Nevada, Reno. She raised two children and now delights in four grandchildren, together with her husband of over fifty years. When her niece died suddenly at thirty-six, Dr. Liebman became a medical genetics detective and an advocate for genetic testing.